AF302721

Fundamental Limitations in the Measurement and Stabilization of the Carrier-Envelope Phase of Ultrashort Laser Pulses

DISSERTATION

zur Erlangung des akademischen Grades

doctor rerum naturalium
(Dr. rer. nat.)
im Fach Physik

eingereicht an der
Mathematisch-Naturwissenschaftlichen
Fakultät I
Humboldt-Universität zu Berlin

von
Dipl.-Phys. Sebastian Koke
geboren am 13.01.1982 in Detmold

Präsident der Humboldt-Universität zu Berlin:
Prof. Dr. Jan-Hendrik Olbertz

Dekan der Mathematisch-Naturwissenschaftlichen
Fakultät I:
Prof. Dr. Andreas Herrmann

Gutachter:
1. Prof. Dr. Thomas Elsässer
2. Prof. Achim Peters, Ph.D.
3. Prof. Erich Ippen, Ph.D.

Tag der mündlichen Prüfung: 24.10.2011

Bibliografische Information der Deutschen Nationalbibliothek
Die Deutsche Nationalbibliothek verzeichnet diese Publikation in der Deutschen
Nationalbibliografie; detaillierte bibliografische Daten sind im Internet über
http://dnb.d-nb.de abrufbar.
1. Aufl. - Göttingen: Cuvillier, 2012
Zugl.: Berlin, Univ., Diss., 2011

978-3-95404-006-3

© CUVILLIER VERLAG, Göttingen 2012
Nonnenstieg 8, 37075 Göttingen
Telefon: 0551-54724-0
Telefax: 0551-54724-21
www.cuvillier.de

Abstract

The stabilization of the carrier-envelope phase of ultrashort laser pulses went through a rapid development from the first publication of a feasible concept in 1999 to being a mature tool for frequency metrology and attosecond science now. Using this technique, stabilization of the timing between the carrier wave and the envelope of a laser pulse with residual jitters of only 100 attoseconds has become possible. Naturally, the questions arises whether and how this can be further improved.

The current work is devoted to determining the physical mechanisms which generate jitter in carrier-envelope phase stabilization. Furthermore, it is investigated whether there is a fundamental limitation to the achievable accuracy.

To this end, two methods for removal of technical noise contributions are initially discussed. Different interferometer topologies are investigated and spurious interferometer noise is reduced by more than 40% using a common-path layout. A novel two-detector based carrier-envelope phase retrieval technique for amplified laser pulses is demonstrated enabling the circumvention of the shot-noise constraint of the conventional extraction method to the maximum extent possible. Next, a novel feed-forward stabilization concept is developed that enables carrier-envelope phase stabilizations with only 20 attosecond residual timing jitter between carrier and envelope of the laser pulse. This feed-forward method is unconditionally stable against drop-out and permits the generation of a train of pulses with identical electric field structure with no additional measures. As the feed-forward concept widely avoids the technical noise sources of the conventional feedback stabilization, the resulting noise spectra exhibit only two unavoidable residual noise mechanisms: a high-frequency white noise floor stemming from shot noise in the carrier-envelope phase detection and a drift-like contribution with $1/f$ noise characteristics.

Finally, the drift-like residual noise mechanism is found to induce phase noise below the level expected for the conversion of pump laser shot noise into carrier-envelope phase jitter. A feedback based squeezing, a photon-number squeezing process and a quantum non-demolition like conversion are discussed as possible explanations for this striking finding. It is shown that either the feedback squeezing or the quantum non-demolition process is the probable origin for the observed sub-shot-noise signatures of the carrier-envelope phase jitter.

Zusammenfassung

Die Stabilisierung der Carrier-Envelope-Phase ultrakurzer Laserimpulse weist eine rapide Entwicklung von der ersten Publikation eines Konzeptes bis hin zu dem heute vorhandenen ausgereiften Werkzeug für die Frequenzmetrologie und Attosekunden-Impulserzeugung auf. Mit Hilfe dieser Technik ist die Stabilisierung des Zeitbezugs zwischen der Trägerwelle und der Einhüllenden eines Laserimpulses mit einem Restjitter von nur 100 Attosekunden erreicht worden. Naturgemäß tritt die Frage auf, ob und wie diese Resultate weiter verbessert werden können.

Die vorliegende Arbeit ist dem Auffinden der Quellen des Restrauschens gewidmet. Darüber hinaus wird untersucht, ob die Genauigkeit der Carrier-Envelope-Phasenstabilisierung grundsätzlichen Beschränkungen unterliegt.

In diesem Rahmen, werden zuerst zwei Verfahren zur Verhinderung von technischen Rauschbeiträgen diskutiert. Verschiedene Interferometertopologien werden miteinander verglichen, und es wird gezeigt, dass eine Auslegung, in der die Interferometerarme möglichst deckungsgleich verlaufen, eine Reduktion künstlicher Rauschbeiträge um mehr als 40% ermöglicht. Zusätzlich wird ein neuartiges Verfahren zur Extraktion der Carrier-Envelope-Phase von verstärkten Laserpulsen mittels nur zweier Detektoren demonstriert, das die Schrotrauschbegrenzung der konventionellen Extraktionsmethode deutlich reduziert. Danach wird ein Feed-Forward Konzept zur Carrier-Envelope-Phasenstabilisierung entwickelt und für die Begrenzung der Fluktuationen auf nur 20 Attosekunden verwendet. Diese Feed-Foward Methode kann nicht ausfallen und erlaubt die einfache Erzeugung eines Impulszuges bestehend aus Impulsen mit identischem elektrischen Feld. Aufgrund der weitgehenden Vermeidung der technischen Rauschquellen des konventionellen Stabilisierungsverfahrens, weisen die mit dem Feed-Forward Verfahren erzielten Rauschspektren nur zwei Beiträge auf: einen weißen Rauschteppich für hohe Frequenzen und ein driftartiges Ansteigen zu niedrigen Frequenzen hin.

Abschließend wird gezeigt, dass die driftartigen Beiträge zu einem Phasenrauschen korrespondieren, das unterhalb des Niveaus liegt, welches für die Konversion von Schrotrauschen der Pumpquelle in Carrier-Envelope-Phasenrauschen erwartet wird. Die Erzeugung von gequetschtem Licht durch den Rückkoppelprozess und durch eine nichtlineare optische Wechselwirkung sowie eine zerstörungsfreie Quantenmessung werden als mögliche Erklärungsansätze diskutiert. Es wird gezeigt, dass entweder der Rückkoppelprozess oder die zerstörungsfreie Quantenmessung die Ursache des nicht durch das Schrotrauschen begrenzten Restjitters ist.

Contents

Introduction

In the last 50 years, laser science has progressed from the first demonstration of a laser [1, 2] to a research field that has tremendous impact on other fields in science as well as our everyday life. Utilization of the unique properties of laser light is diverse and ranges from highly sophisticated tools in experimental physics to core components in contemporary telecommunication networks. Such applications are the driving force behind the continuous progress of laser science and have initiated the development of novel laser concepts.

Apart from laser sources emitting a continuous wave, the generation of pulsed laser light has attracted much attention. To a large part, this interest is due to the fact that laser pulses enable the study of rapid dynamic processes in nature that cannot be accessed electronically. For example, the dynamics of chemical bonds during the course of a chemical reaction can be temporally resolved with femtosecond laser pulses (1 fs = 10^{-15} s) [3]. The strive for enhancement of the temporal resolution in such studies has led to the invention of the technique of mode locking [4, 5]. The basic concept of mode locking is phase-synchronization of laser cavity modes, which are equally spaced in frequency. While originally in competition with other methods for the generation of laser pulses, mode locking has succeeded in the generation of the shortest laser pulses. This success has culminated in the production of only 4.5 fs long laser pulses directly from a laser [6]. These laser pulses encompass less than two optical cycles in their intensity half width.

With ever decreasing pulse duration, the traditional description of a pulse via its slowly-varying intensity envelope breaks down as the relative phase between carrier and the maximum of the intensity envelope starts to play a role. This quantity is called the carrier-envelope phase. The breakdown of the slowly-varying envelope approximation appears most markedly in high-harmonic processes [7, 8] and is utilized for the synthesis of attosecond pulses in the extreme ultraviolet (1 as = 10^{-18} s) [9]. Since there is no mechanism locking the carrier-envelope phase of a pulse in the generation process, the carrier-envelope phase evolves freely in a train of pulses from a mode-locked laser. The evolution is dominated by a constant drift, with deviations only caused by external perturbations of the mode-locking process. In the frequency domain, the carrier-envelope phase drift corresponds to an offset of the comb of the synchronized cavity modes from zero. While this fact was already realized in the late 70s [10], there was no practical way of controlling the offset of the frequency comb of a mode-locked laser, i.e., the electric field evolution in a pulse train, until much later. For the first time a self-referencing measurement technique for the comb offset, termed f–$2f$ interferometry, was demonstrated in 1999 [11, 12]. The availability of this measurement scheme has revolutionized the field of frequency metrology [13, 14], where broadband frequency combs produced by few-cycle lasers are now widely used as optical clockwork [14, 15] replacing the formerly used complicated and inflexible frequency chains [16]. Frequency comb based metrology setups have gained such a high degree of precision that the question of a potential drift of fundamental constants can be addressed today [17].

Both attosecond physics and frequency metrology require a controlled evolution of the carrier-envelope phase in a pulse train. Currently, the most widespread technique for achieving this goal is phase locking of the comb offset to a stable reference using a phase-locked loop. Feedback is applied to the laser by manipulating the difference between the intracavity group and phase velocity. Most often this is achieved by modulation of the pump of the mode-locked laser, which alters the nonlinear pulse propagation inside the gain medium. The residual jitters between carrier and envelope that have been achieved with such phase-locked loops are on the order of only 100 as.

In order to progress beyond current achievements of carrier-envelope phase stabilization, however, clarification of the origin of the residual carrier-envelope phase jitter is required. The work presented in this thesis investigates whether the origin of the residual jitter is merely of a technical nature or if there is a fundamental physical limitation.

Outline

After a general introduction presented in Chapter 1, improvements of the existing technology are devised in Chapter 2. It is shown that a common-path layout of f–$2f$ interferometers removes spurious technical noise by more than 40%. In addition, a two detector arrangement for carrier-envelope phase retrieval of amplified laser pulses reducing the impact of shot noise is demonstrated. Both investigations give insight into mechanisms that currently limit carrier-envelope phase stabilization. In Chapter 3, a novel approach to stabilization is demonstrated, which replaces the feedback concept by a feed-forward scheme. In contrast to feedback based carrier-envelope phase stabilization, the feed-forward scheme is capable of generating zero-offset frequency combs, is unconditionally stable and does not require any manipulation of important laser parameters. Moreover, with the achieved 20 as residual temporal jitter between carrier and envelope, the feed-forward carrier-envelope phase stabilization clearly outperforms the feedback method promising to be a substantial leap ahead for frequency metrology and attosecond physics. Finally in Chapter 4, analysis of the obtained stabilization results reveals that in Kerr-lens mode-locked lasers the residual carrier-envelope phase jitter lies below the level expected for the conversion of pump laser shot noise into carrier-envelope phase noise. In search for an explanation for this remarkable finding, a feedback based squeezing, a soliton induced photon-number squeezing and a quantum non-demolition like light power measurement is discussed. The feedback based squeezing and the quantum non-demolition like measurement of the intracavity power are identified as the probable origins of the observed sub-shot-noise signatures of the carrier-envelope phase jitter. Application of this sub-shot-noise sensitivity may enable the limitations in current high-sensitivity interferometric measurements to be overcome.

Chapter 1

Methods and Concepts

In the following chapter, the scientific background for the work presented in this thesis is introduced and reviewed. While the first two sections deal with the mode-locking mechanism and the definition of the carrier-envelope phase (CEP) the third section covers the measurement setups used to measure the CEP drift. Then the conventional feedback-loop for stabilization of the CEP drift is introduced, and the supercontinuum generation process employed in measurement setups is discussed. Finally, different CEP-noise generating mechanisms are detailed.

1.1. Mode-Locked Lasers

Among the techniques for the generation of laser pulses, mode locking enables the generation of the shortest laser pulses. It was demonstrated in 1964 for the first time [4, 5], and the unique feature that sets mode locking apart from other pulse generating methods is that it bases on the establishment of a fixed phase relationship between several longitudinal modes [18]

$$\nu_m = \frac{mc}{2nL} \tag{1.1}$$

of the resonant laser cavity. Here, c is the vacuum speed of light, n is the refractive index of the medium inside the cavity, and L denotes length of the cavity. In analogy to the emergence of a beat from the superposition of two oscillations with discrete frequencies, the interference of the larger number of contributing modes in a mode-locked laser results in the periodic production of light bursts at instances of constructive interference and a vanishing electric field in between those bursts. The temporal separation T_R of these light bursts is given by the inverse of the mode spacing $\delta\nu = c/2nL$ and the temporal confinement of them increases the larger the

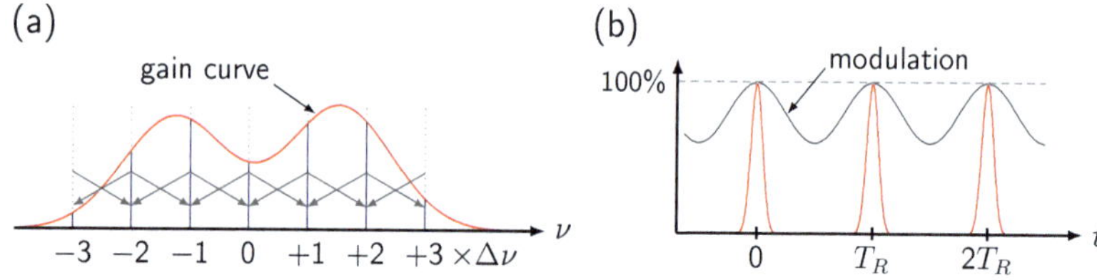

Figure 1.1.: Frequency domain (a) and time domain (b) representation of active mode locking. The imposed amplitude modulation generates sidebands to the initial oscillation mode labeled with 0 at a frequency offset that is given by the modulation frequency. If the modulation frequency matches the laser cavity mode spacing, successive sideband generation creates a frequency comb structure encompassing the gain bandwidth. By the mutual coupling of the comb modes the relative phase is fixed such that a pulse train is observed in the time domain. The pulses pass the modulator at times of maximum transmission.

number of coupled modes is. For example, if $2M + 1$ modes of electric field strength $\mathcal{E}_0$ having no phase difference superpose, the electric field is [18]

$$\mathcal{E}(t) = \sum_{m=-M}^{M} \mathcal{E}_0 \exp[i2\pi(\nu_c + m\,\delta\nu)\,t]$$
$$= \mathcal{E}_0 \frac{\sin\left[(2M+1)\,\delta\nu\,t/2\right]}{\sin(\delta\nu\,t/2)}\,\exp(i2\pi\nu_c t) \tag{1.2}$$

(ν_c carrier frequency). Hence, the pulse duration is

$$\tau \approx \frac{1}{(2M+1)\,\delta\nu} = \frac{1}{\Delta\nu} \tag{1.3}$$

and decreases the larger the spectral width $\Delta\nu = (2M+1)\,\delta\nu$ of contributing modes is.

Mode locking can be achieved by active or passive means. In active mode locking, a modulator, which is placed inside the cavity, induces either an amplitude or phase modulation. In the former case, for example, the active variation of the amplitude induces frequency sidebands on the light passing through the modulator. The frequency offset of these sidebands is given by the modulation frequency f_m. If the modulation frequency is equal to the mode spacing of the laser cavity ($f_\mathrm{m} = \delta\nu$), the newly generated frequencies fall on top of cavity modes and can, thus, propagate without being subject to destructive interference. Furthermore, if these modes still fall within the gain bandwidth of the laser, they are amplified and can act as seed for the generation of additional sidebands. As illustrated in Fig. 1.1(a), eventually the complete gain bandwidth is covered with cavity modes exhibiting a fixed phase relationship, which is imposed due to the successive seeding from sidebands of the other modes. In the time domain [see Fig. 1.1(b)], this procedure corresponds to generation of pulses that pass the modulator after each roundtrip at times of transparency.

In passive mode locking, such an amplitude modulation is induced by the pulse itself in an element offering saturable absorption (*self-amplitude modulation*). The saturable absorber favors the high intensities of light bursts created by the superposition of initially unlocked oscillation modes. The largest of these bursts suffers the least attenuation such that it grows more rapidly than the others in subsequent cavity roundtrips. Eventually, a stable intense pulse forms out by the balance between saturable absorption induced pulse shortening and pulse broadening arising from the finite gain bandwidth [18]. In the first demonstration of passive mode locking, saturable absorption has been accomplished by incorporating a dye solution in a flash-lamp pumped laser [19]. These dyes bleach out for high pulse intensities by the depopulation of dye molecules in the groundstate. Currently, implementation of saturable absorption relying on the excitation of valence band electrons to the conduction band of semiconductors [20] or carbon nanotubes [21] is an active field of research. Compared to dye solutions, these saturable absorbers excel due to their ease of use and due to a faster relaxation. While dyes offer an upper state relaxation time of several tens of picoseconds to nanoseconds, the fastest recombination channel in semiconductor saturable absorbers typically has a relaxation time of 100 fs and originates from the intraband thermalization of conduction band electrons. In addition, these semiconductor saturable absorbers show a relaxation of the absorption on picosecond and nanosecond time scales that is induced by intraband thermalization of conduction band electrons with the lattice and interband recombination [18].

Depending on the ratio of pulse duration τ and saturable absorber relaxation time τ_A one distinguishes between the regimes of *fast saturable absorber mode locking* ($\tau_A \ll \tau$) and *slow saturable absorber mode locking* ($\tau \ll \tau_A$) [22, 23]. In the latter case, generation of pulses shorter than the absorber relaxation time is enabled by gain saturation confining the temporal window of net positive gain below the relaxation time of the absorber [23].

Fast saturable absorber mode locking using real absorbers is not widely used for the generation of femtosecond pulses due to the lack of rapidly relaxing media. However, this passive mode locking regime is accessed when reactive nonlinearities are used to mimic saturable absorption. As no excitation of electrons is involved, such nonlinearities are intrinsically fast and can act to produce a self-amplitude modulation down to the few femtosecond regime. Nonlinear refraction finds widespread application in mode locking of fiber laser by, e.g., nonlinear polarization rotation [24, 25] and in techniques for mode locking of bulk solid-state lasers such as *additive pulse mode locking* [26, 27] and *Kerr-lens mode-locking (KLM)*

[28]. In KLM, the nonlinear propagation of the laser pulses inside the gain medium leads to self-focusing, which is subsequently translated into an amplitude modulation by the higher transmission through some aperture inside the laser cavity [29, 30]. This aperture may result from the mirror dimensions, the diameter of the amplifying medium, or any other aperture intentionally placed inside the cavity. As the self-focusing is the stronger the higher the peak intensity of the laser pulse is, the combined effect of self-focusing and aperturing effectively results in a saturable absorption. The KLM technique is widely used for mode locking of Ti:sapphire lasers and enabled the production of < 4.5 fs long laser pulses [6], which are the shortest pulses obtained directly from an oscillator.

The pulse formation and pulse shaping inside KLM lasers is understood in the framework [31] that has been developed for the description of additive pulse mode locking. According to this theory, the saturable absorber action of self-focusing and aperturing is necessary for the initial formation of pulses as well as their stabilization against growth of noise in between the pulses. The pulse shaping mechanism in steady-state mode locking, however, is due to the interplay of group-delay dispersion (GDD) and self-phase modulation (SPM). For the important case of net negative GDD, a dispersion-managed solitary pulse forms by the balance between intracavity GDD and SPM producing self-stabilized pulse shapes after each roundtrip [32, 33]. To the end of generating short pulses having broadband spectra, it is necessary to carefully control the intracavity GDD. A major breakthrough in this respect has been the fabrication of chirped mirrors [34]. These mirrors consist of a stack of dielectrics with varying thickness, such that the the penetration depth of different wavelengths into the mirror stack varies. Thus, in contrast to the established prism or grating compressors, engineering of the thickness and composition of the layers permits compensation of higher order GDD terms, which is necessary for accessing the few-cycle regime.

1.2. The Carrier-Envelope Phase

In the previous section, it has been elucidated that mode locking and the generation of laser pulses are equivalent to the creation of a comb in the frequency domain. By the example of active mode locking it has been shown that the modes of such a frequency comb are given by multiples of the mode spacing $\delta\nu$. This explanation, however, has neglected the dispersion of the intracavity elements. In general, dispersion has to be accounted for, which has important implications. Due to the dispersion, the group velocity, at which the envelope propagates, is different from the phase velocity, at which the electric field carrier propagates. Thus, the exact electric field structure, which is the product of the carrier and envelope, does not reproduce exactly every pulse. As exemplified in Fig. 1.2, there is a timing difference between the maximum of the envelope and the maximum of the carrier. This tim-

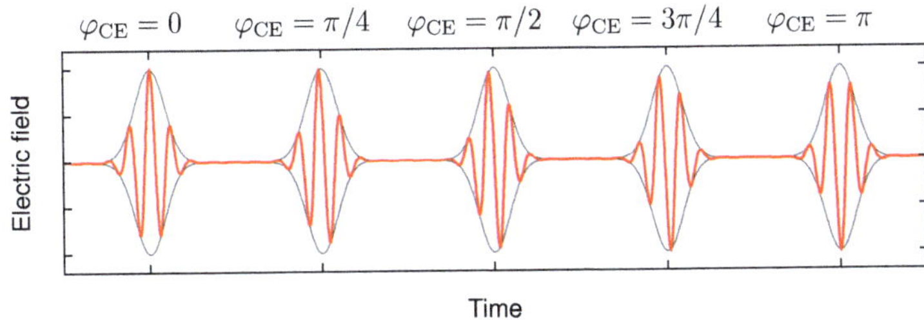

Figure 1.2.: Evolution of the carrier-envelope offset during a pulse train with a difference in carrier frequency and repetition rate adjusted such that the original pulse reproduces itself every eighth roundtrip ($f_{\mathrm{CE}} = f_{\mathrm{rep}}/8$). The envelope of the pulses is shown in gray.

ing difference is conveniently described by the carrier phase difference φ_{CE}, called CEP [1], which corresponds to the temporal offset.

Using this definition of the CEP, the electric field structure of the pulse train may be written as [35]

$$\mathcal{E}(t) = \left[\boldsymbol{A}(t)e^{i\omega_c t} \otimes \sum_{m=-\infty}^{\infty} \delta(t - mT_R) \right] e^{i\varphi_{\mathrm{CE}}(t)} + c.c.\,, \qquad (1.4)$$

where $\omega_c = 2\pi\nu_c$ is the carrier's angular frequency, $\boldsymbol{A}$ is the envelope of a single pulse, and the conventional separation in an envelope $\boldsymbol{A}$ and rapidly evolving electric field oscillations has been applied [2]. In Eq. (1.4), the bracket term consisting of the convolution of a single pulse electric field with the comb of Dirac's delta functions δ represents the periodical production of equal pulses, while the exponential term behind the bracket describes the temporal evolution of the change of phase between carrier and envelope.

To first order [3], this CEP drift is due to the intracavity discrepancy between the group velocity v_g and the phase velocity v_p [11, 36, 37]. Since [35]

$$\frac{1}{v_g} = \frac{1}{c}\left[\frac{dn(\omega)}{d\omega}\omega + n(\omega)\right]_{\omega=\omega_c} = \frac{1}{v_p} + \frac{\omega_c}{c}\left.\frac{dn(\omega)}{d\omega}\right|_{\omega=\omega_c}\,, \qquad (1.5)$$

---

[1] The carrier-envelope phase is sometimes also called *absolute phase*. In the following the term carrier-envelope phase will be used consistently.

[2] The spatial coordinates are neglected for simplicity.

[3] A closer look reveals that there are also contributions of nonlinearities, which are neglected here but will be discussed in Section 1.6 in more detail.

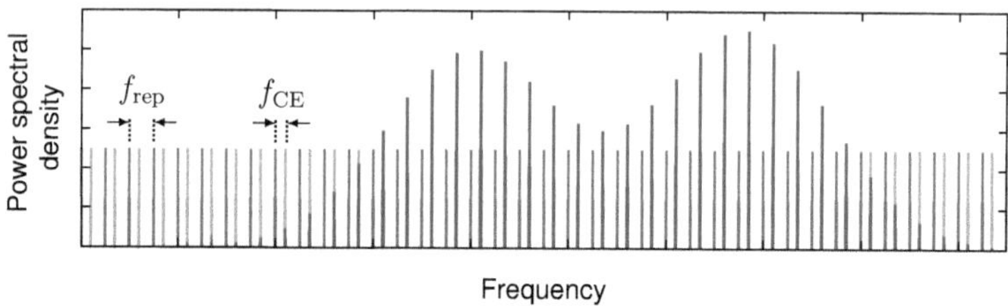

Figure 1.3.: Schematic representation of the frequency comb produced by a mode-locked laser with $f_{\mathrm{CE}} \neq 0$ (red) compared to the frequency comb obtained with $f_{\mathrm{CE}} = 0$ (gray). The two combs are displaced with respect to each other by the carrier-envelope frequency f_{CE}. Note that it is impossible to measure f_{CE} by directly detecting comb mode next to zero since there is no energy contained in it.

the group-phase offset $\Delta\varphi_{\mathrm{GPO}}$ accumulated per-roundtrip due to first-order intra-cavity dispersion is given by

$$\Delta\varphi_{\mathrm{GPO}} = -\omega_c \int_0^L \left(\frac{1}{v_g} - \frac{1}{v_p} \right) dz = \frac{\omega_c^2}{c} \int_0^L \left. \frac{dn(\omega, T, z)}{d\omega} \right|_{\omega=\omega_c} dz \,, \tag{1.6}$$

where c denotes the vacuum speed of light, L is the cavity length, and T denotes the temperature. Typical group-phase offset values for Ti:sapphire lasers are on the order of a few 100 cycles [11]. However, full cycle changes do not alter the electric field structure in the subsequent laser pulse such that only the fractional contribution results in a change of the CEP φ_{CE} per roundtrip

$$\Delta\varphi_{\mathrm{CE}} = \Delta\varphi_{\mathrm{GPO}} \mod 2\pi \,. \tag{1.7}$$

This constant CEP drift leads to a linear change of the absolute CEP value $\varphi_{\mathrm{CE}} = 2\pi f_{\mathrm{CE}} t$, where the carrier-envelope frequency (CEF) f_{CE} is determined by

$$f_{\mathrm{CE}} = \frac{1}{2\pi} \frac{\Delta\varphi_{\mathrm{CE}}}{T_R} \,. \tag{1.8}$$

Apart from the temporal electric field evolution, the CEP drift also affects the frequency domain representation of the pulse train. Fourier transformation of Eq. (1.4) yields the frequency domain representation

$$\tilde{\mathcal{E}}(\nu) = \tilde{\mathbf{A}}(\nu - \nu_c - f_{\mathrm{CE}}) \sum_{m=-\infty}^{\infty} \delta(\nu - m f_{\mathrm{rep}} - f_{\mathrm{CE}}) \,. \tag{1.9}$$

This is a frequency comb that is offset from zero by the CEF f_{CE}, as illustrated in Fig. 1.3. The envelope's spectrum can be written in Euler's representation of complex numbers: $\tilde{\boldsymbol{A}}(\nu) = \left|\tilde{\boldsymbol{A}}(\nu)\right| e^{i\phi(\nu)}$. In the following section, the spectral phase $\phi(\nu)$ will be decomposed into a constant part given by the CEP, i.e., $\varphi_{CE} = \phi(\nu_c)$, and a deviation $\Delta\phi(\nu)$ from constancy: $\phi(\nu) = \varphi_{CE} + \Delta\phi(\nu)$. Applying this decomposition, it is easily seen that the periodic reproduction of pulses with identical pulse envelope but changing CEP corresponds to spectra with the same amplitude and relative phase but different phase offsets in the Fourier domain.

From Eq. (1.9), it follows that all comb lines of a mode-locked laser are determined by just two parameters: the CEF, determining the offset of the comb, and the repetition rate of the laser, which defines the spacing of adjacent comb modes

$$\nu_m = m\, f_{rep} + f_{CE}, \qquad m \in \mathbb{N}. \tag{1.10}$$

In fact, the uncertainty of comb mode equidistancy has also been experimentally checked to be below 10^{-15} [38, 39]. Making use of this low uncertainty of frequency combs, optical frequency synthesis and comparison with an uncertainty approaching 10^{-19} has been demonstrated [40]. This renders frequency combs indispensable tools in contemporary high-precision frequency metrology experiments [13, 41], such that, e.g., the exploration of tiny drifts of fundamental constants is within reach [17].

The influence of the CEP on the temporal electric field evolution, in contrast, permits the generation of isolated attosecond pulses [9], which consist of higher harmonics of the driving electric field's carrier frequency. The higher harmonics are emitted every half cycle when the electric field of the driver pulse has zero crossings. At these points in time, the kinetic energy the electrons gained during the laser driven acceleration can be emitted as high-energy photons (≈ 100 eV) in the recollision with the parent ion [42]. Furthermore, the requirement of constructive interference between electron trajectories ionized at different instances of time limits the duration of each individual high energy photon burst to a fraction of a half-cycle of the driving electric field. For the case of a Ti:sapphire laser pulses with a carrier frequency of $\lambda = 800$ nm, one period of the electric field corresponds to a time span of ≈ 2.5 fs, such that a fraction of a half-cycle is in the attosecond regime. As a result of these temporal characteristics of high-harmonic radiation, production of isolated attosecond pulses only requires selection of one of those high energy photon bursts. Such a selection can either be achieved by energy filtering of the radiation of the particular half-cycle electron burst with the maximum kinetic energy [9, 43, 44] or by an additional gating of the ionization, utilizing the interference between two driver pulses [45, 46]. Both methods, however, require a driving electric field with CEP $\varphi_{CE} = 0$. Therefore, generation of attosecond pulses crucially depends on the CEP and the quality of its stabilization.

With the help of the correspondence between the frequency domain and time domain representation of the CEP developed in this section, the question of how to measure the CEF f_{CE} as well as the CEP φ_{CE} can now be addressed in the next sections.

1.3. Measuring the Carrier-Envelope Phase Evolution

The characterization of properties of ultrashort pulses is important for the understanding of the physical effects studied with these pulses. The general difficulty of such a characterization is that these pulses are the shortest man-made events. Hence, ultrashort pulses cannot be sampled by any shorter controllable event. As a result, ultrashort pulse characterizations typically measure the pulse properties by employing an interaction of the pulse under test with itself. Apart from the ubiquitous autocorrelation technique, more advanced pulse self-characterization techniques like frequency resolved optical gating (FROG) [47] and spectral phase interferometry for direct electric-field reconstruction (SPIDER) [48] have been developed. Unfortunately, neither of these methods is capable of measuring the CEP due to their insensitivity towards changes of the CEP φ_{CE} [49].

In 1996 Xu et al. demonstrated a technique capable of detecting the CEP drift rate [50] for the first time. The method they proposed is based on analyzing the cross-correlation between two successive laser pulses. As the electric field structure changes if a CEP drift is present, the symmetry of the cross-correlation is distorted, and the magnitude of this distortion gives information about the magnitude of the drift rate. However, considerable experimental efforts are necessary since the measured CEP drift is easily corrupted by the dispersion of air in the arms of the interferometer. This issue mandates operation of the cross-correlator in vacuum. As a result of the complexity of such a setup and due its inapplicability for a phase-lock [11, 51], the cross-correlation method has not found widespread application.

Instead, analyzing the interference between different harmonics of the pulse spectrum is routinely used today [11, 12]. These nonlinear interferometers are based on the following fact [52]. During a nonlinear frequency conversion, the CEP of the interacting waves add up like their frequencies. For example, if the CEP of the m-th pulse in a pulse train is denoted by φ_m, then the spectrum of the m-th pulse after frequency doubling is determined by

$$
\begin{aligned}
\tilde{\boldsymbol{\mathcal{E}}}^{\mathrm{SH}}(\nu) &\propto \int \chi^{(2)}(\nu; \nu', \nu - \nu') \tilde{\boldsymbol{A}}(\nu) \tilde{\boldsymbol{A}}(\nu - \nu') d\nu' \\
&\propto e^{i 2\varphi_m} \int \chi^{(2)}(\nu; \nu', \nu - \nu') |\tilde{\boldsymbol{A}}(\nu)| |\tilde{\boldsymbol{A}}(\nu - \nu')| e^{i(\Delta\phi(\nu') + \Delta\phi(\nu - \nu'))} d\nu' \quad (1.11) \\
&\equiv |\tilde{\boldsymbol{A}}^{\mathrm{SH}}(\nu)| e^{i(2\varphi_m + \Delta\phi^{\mathrm{SH}}(\nu))},
\end{aligned}
$$

where $\chi^{(2)}$ is the second-order nonlinear susceptibility. Thus, the phase of the second harmonic is

$$\phi_m^{\mathrm{SH}}(\nu) = 2\varphi_m + \Delta\phi^{\mathrm{SH}}(\nu)\,. \tag{1.12}$$

Equation (1.12) provides two insights. First, under the assumption of a constant CEP drift rate $\varphi_m = m\Delta\varphi_{\mathrm{CE}}$, it is easily derived that the CEP drift rate of the frequency doubled pulse train is twice the drift rate of the original pulse train, i.e., $\phi_{m+1}^{\mathrm{SH}} - \phi_m^{\mathrm{SH}} = 2\Delta\varphi_{\mathrm{CE}}$. Second, the quantity φ_m determines whether the original pulse train and the frequency-doubled pulse train – or, in general, odd and even harmonics of the pulse train – interfere constructively or destructively at the output of the nonlinear interferometer [11]. Hence, analysis of such an interference gives access to the CEP.

The nonlinear interferometer that found most widespread application for CEP retrieval is the f–$2f$ interferometer [11, 12, 53], but also f–0 [54] and $2f$–$3f$ interferometers [55] have been utilized[4]. All these implementations require a sufficiently broad incident spectrum. If the spectrum of the pulse under study lacks this criterion, spectral broadening in micro-structured fibers [56] or bulk media [57] has to be employed before the beam enters the nonlinear interferometer. A detailed discussion about this supercontinuum generation step will be given in Section 1.5.

While the principle of nonlinear interferometry is generally applicable to every laser pulse source, the way how the interference is read out has to be adapted to the laser source.

f–$2f$ Interferometry for Oscillator Pulses

For ultrafast laser oscillators, with repetition rates typically in the MHz or GHz range, the beat between the frequency combs of the fundamental and second harmonic light is analyzed by *radio-frequency (RF) heterodyning* [11, 12]. This technique is summarized in Fig. 1.4. The beat between the second harmonic of a comb mode ν_n and the adjacent fundamental comb mode ν_{2n} manifests itself in an amplitude modulation of the pulse train detected at the output of the f–$2f$ interferometer. A typical setup of such an f–$2f$ interferometer is depicted in Fig. 1.5(a) and has been experimentally realized in the year 2000 for the first time [58]. Figures 1.5(b) and (c) show the amplitude modulated pulse train, which is measured by the avalanche photodiode (APD). Using Eq. (1.10), the beat frequency is calculated to be the CEF

$$f_{\mathrm{beat}} = 2\nu_n - \nu_{2n} = 2nf_{\mathrm{rep}} + 2f_{\mathrm{CE}} - (2nf_{\mathrm{rep}} + f_{\mathrm{CE}}) = f_{\mathrm{CE}}\,. \tag{1.13}$$

---

[4]As only f–$2f$ interferometer are used throughout the thesis the following presentation concentrates on this configuration. However, most of what is discussed is also valid for the other configurations, such that every mentioning of f–$2f$ can be regarded as a synonym for this class of nonlinear interferometry.

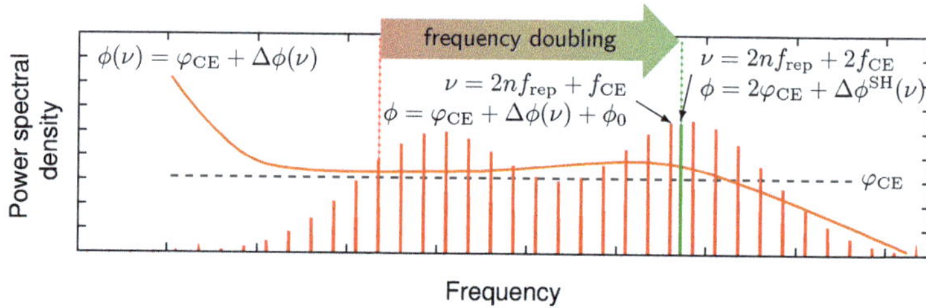

Figure 1.4.: Principle of f–$2f$ interferometery. The low frequency part of a frequency comb (red) is frequency doubled and heterodyned with comb modes in the high frequency region. The phase of the frequency doubled light is determined by twice the spectral phase (orange) at the fundamental frequency plus a constant offset [52]. This makes the interference sensitive to changes of the carrier-envelope phase.

Furthermore, the temporal phasing of the beat is determined by the CEP since the two interfering combs have a φ_{CE}-dependent phase difference [see Fig. 1.4 and Eq. (1.12)]. This has an important consequence. While a phase deviation may result from integration of tiny frequency mismatches in techniques stabilizing the CEF, such as the technique developed by Xu et al. [50], these artifacts are absent in stabilizations that utilize a CEP dependent error signal like RF heterodyning. However, the true value of the pulse's CEP is not retrievable from the temporal phasing of the beat since the possibly non-zero interferometer arm length difference introduces an unknown contribution ϕ_0 to the temporal phasing of the beat.

f–$2f$ Interferometry for Amplifier Pulses

Due to $1/f$ like noise contributions, the RF heterodyning technique is not applicable for amplifiers which typically have lower repetition rates in the few kHz range. Instead, spectral interference can be utilized to detect the CEP evolution [53, 59]. To this end, the fundamental and frequency doubled pulses are temporally delayed with respect to each other and inserted into a spectrograph. Due to the temporal delay, a fringe pattern is observable in the spectrum

$$\begin{aligned}
S(\nu) &\propto \Big| |A(\nu)|e^{i\phi(\nu)+\phi_0} + |A_{\mathrm{SHG}}(\nu)|e^{i(\pi/2+2\phi(\nu/2)+2\pi\nu\tau_d)} \Big|^2 \\
&= |A(\nu)|^2 + |A_{\mathrm{SHG}}(\nu)|^2 + 2|A(\nu)||A_{\mathrm{SHG}}(\nu)|\times \\
&\quad \times \cos\Big(2\pi\nu\tau_d + \varphi_{\mathrm{CE}} + \Delta\phi^{\mathrm{SH}}(\nu) - \Delta\phi(\nu) + \phi_0\Big),
\end{aligned} \qquad (1.14)$$

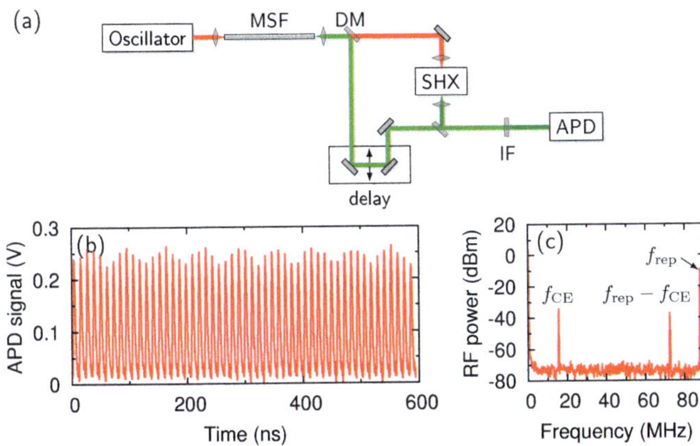

Figure 1.5.: (a) Example of an f–$2f$ interferometer in Mach-Zehnder configuration. In this example, the output of the oscillator is spectrally broadened in a micro-structured fiber (MSF) and split into short and long wavelength contributions by means of a dichroic mirror (DM). The long wavelength contributions are frequency doubled in a second harmonic crystal (SHX), and after combination of the two interferometer arms the beams are spectrally filtered by an interference filter (IF) before being detected by an avalanche photodiode (APD). (b) Experimental example trace showing the amplitude modulation of the pulse train detected by a diode at the output of the f–$2f$ interferometer. (b) Corresponding radio-frequency analyzer trace with the beat note f_{CE}, the mirror beat note $f_{\mathrm{rep}} - f_{\mathrm{CE}}$, and the first intermode beat note f_{rep}.

where τ_d is the temporal delay. Hence, the phasing of the fringe pattern shifts with variations of the CEP. Yet again, the unknown constant ϕ_0 prohibits the extraction of the φ_{CE} of a single pulse.

In principle, the spectral interference based scheme is also capable of single-shot CEP measurements of pulse trains with repetition rates higher than a few kHz. However, the conventional way of extracting the position of the fringes via recording with a CCD camera sets an upper limit for feasible repetition rates. In Section 2.2, a method that relaxes this restriction will be devised.

Determination of the Carrier-Envelope Phase of a Single Pulse

Beyond $f\text{--}2f$ interferometry, several methods have been developed for the determination of the CEP of a single pulse. In order to exclude unknown interferometer contributions ϕ_0, these methods rely on the analysis of ionization properties. For example, above-threshold ionization (ATI) of noble gas atoms exhibits an asymmetry that can be read out by two oppositely placed time-of-flight detectors [60]. Alternatively, the CEP may be retrieved via determination of the cut-offs corresponding to individual driving electric field half-cycles in high-harmonic spectra [61] or by analysis of the THz wave emitted from plasmas formed by focused few-cycle pulses [62]. A method accessible with oscillator pulse energies is the detection of the CEP asymmetry of photoemission of electrons from metals [63, 64].

Except for the ATI method [65], these methods require averaging in order to resolve the CEP dependence. This is the reason why the above mentioned techniques have not widely replaced $f\text{--}2f$ interferometry for CEP stabilization yet.

1.4. Stabilization of the Carrier-Envelope Phase

In addition to the regular dispersion-induced changes in the CEP discussed so far, the CEP is subject to environmental influences, which cause the CEP to jitter. Among these environmental influences are dispersion variation due to fluctuations of, e.g., pressure or temperature, fluctuations of the cavity length by mirror vibrations, or fluctuations of the pump laser's intensity. In order to reduce this jitter, a mechanism that is capable of counteracting these environmental influences must be provided. In the following, the conventional feedback based approach is outlined. Despite its proven utility, the feedback based method has distinct drawbacks which will be discussed later in Chapter 3 together with a novel scheme circumventing these drawbacks.

Carrier-Envelope Phase Stabilization of Oscillators

For CEP stabilization of oscillators, a servo loop is typically built as a phase-locked loop (PLL) as shown in Fig. 1.6. First a phase-proportional error signal is generated by a phase comparator. In analog variants of such phase comparators, this is

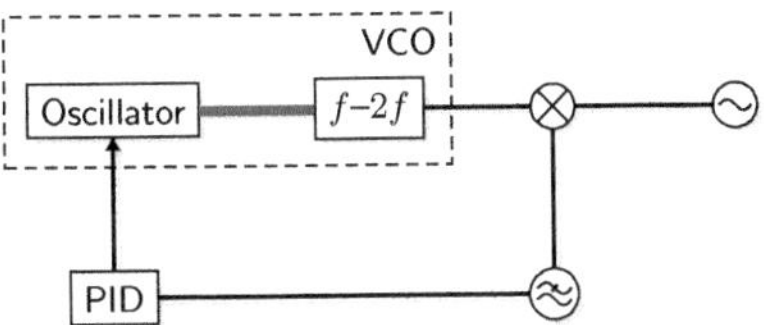

Figure 1.6.: Illustration of an analog phase-locked loop (PLL) based carrier-envelope phase stabilization, as typically applied for oscillators. In this PLL the carrier-envelope frequency is the quantity that is controlled such that the oscillator in combination with the f–$2f$ interferometer resembles the voltage controlled oscillator (VCO) of conventional PLLs.

accomplished by mixing the beat signal with a reference signal and low pass filtering the mixing product. By deriving the reference signal phase coherently from the repetition rate via frequency division, spurious CEP jitter due to drifts of the reference oscillator's frequency relative to the repetition rate of the laser is prevented. Furthermore, the range of unambiguous phase difference retrieval may be expanded beyond the limit of π by replacing the analog phase comparator with a digital one. While such an replacement is accompanied by a reduced sensitivity towards small phase distortions, the increased hold-in range of digital PLLs permits maintaining the stabilization over longer periods of times. After phase comparison, the feedback loop is closed by processing the phase error signal with an amplifier and feeding it back to the laser. While, in principle, a proportional amplification of the error signal already suffices for stabilization, using a proportional-integral-differential (PID) controller for feedback signal generation typically improves the short- and long-term performance of the stabilization.

Since variations of the CEP drift stem from changes in the intracavity dispersion, techniques enabling manipulation of the intracavity dispersion, in turn, allow controlling the CEP in such feedback loop applications. Three different feedback methods found widespread application: shifting of intracavity wedges [66], mechanical rotation of an intracavity prism compressor end mirror [58], and modulation of the pump power by means of an acousto-optical modulator (AOM) [67]. While the first method changes the intracavity dispersion by insertion or reduction of wedge material transversed by the beam, tilting the prism compressor end mirror alters the per-roundtrip group delay and, hence, the group-phase offset [68]. Since both of these methods require mechanical movements, they are limited in speed by the inertia of the elements moved, resulting in demonstrated maximum bandwidths of some 20 kHz [69].

In contrast, the pump power modulation based scheme draws on modulating the intracavity peak power. As a peak power change translates into a CEP modulation via nonlinear pulse propagation, the pump power modulation based schemes, in principle, permit CEP modulations up to the frequency of relaxation oscillations [70]. Usage of an AOM as modulator sets an additional bandwidth constraint resulting from the time lag due to the travel time of the acoustic wave from the transducer to the interaction region. Both constraints allow for CEP modulation bandwidths which are on the order of MHz. The most direct nonlinear pulse propagation mechanism that can explain such a nonlinear CEP shift is the Kerr-induced nonlinear refractive index modulation [71]. This effect alters group and phase velocity differently. However, there are additional mechanisms causing an intracavity peak power into CEP drift conversion which can be more complex and will be discussed in Section 1.6.

The PLL electronics in combination with the chosen feedback method leads to a characteristic amplitude and phase response of the loop [72]. In order to achieve damping of the CEP jitter, the loop gain has to be adjusted such that the phase margin is greater than zero. This phase margin is defined as the difference between π and the phase of the loop's output signal relative to its input at the frequency where the gain crosses unity. Actually, this constraint is what is limiting contemporary AOM based stabilization schemes. Even in investigations where the loop characteristics have been carefully tailored for optimal phase noise suppression, loop bandwidths of only ≈ 100 kHz have been obtained [72]. Thus, any phase noise above 100 kHz observed in those experiments is not corrected [72].

Best reported stabilizations are achieved with oscillators that are octave spanning or close to it [73, 74]. Phase errors of only ≈ 100 mrad integrated from half of the repetition rate of the laser to 0.2 mHz are achieved for such lasers with dominant contributions within the frequency range between 100 Hz and 50 kHz.

Carrier-Envelope Phase Stabilization of Amplifiers

Beyond the CEP stabilization of the oscillators, amplifier systems need a second stabilization loop that compensates for CEP distortions that happened, e.g., within the stretching and compression of the pulses [for chirped pulse amplification (CPA)], pulse picking, and the amplification step itself [75, 76]. Since these distortions typically happen on slower time scales than the CEP jitter in oscillators, this stabilization loop is often called the *slow loop*. The CEP behind the amplifier is conventionally detected with a spectral interference based f–$2f$ interferometer, and the deviation of the fringe position relative to some reference is determined by recording the spectral interference pattern with a CCD camera and processing it with a PC. To simplify matters, the computed error signal is often fed back to the same intracavity dispersive element which is used for the oscillator CEP stabilization. In order to prevent distortions of the oscillator CEP stabilization, elements external to the oscillator such as prism or grating stretchers and compressors [77, 78] or programmable acousto-optic dispersive filters [79] can be used.

The best reported CEP stabilization results lie in the $100 - 300$ mrad range [80, 81]. However, these values are obtained with a detection averaging over multiple laser shots preventing a truthful characterization of the residual CEP jitter. In Section 2.2, a single-shot detection of the CEP behind the amplifier is presented and the jitter is analyzed on a pulse-to-pulse basis for the first time.

1.5. Supercontinuum Generation for f–$2f$ Interferometry

As already mentioned, f–$2f$ interferometry requires an octave spanning spectrum for the overlap between fundamental and frequency-doubled contributions. If the spectrum of the laser source under study does not exhibit such a large bandwidth, the spectrum can be broadened external to the laser via nonlinear interaction. Again, the employed technique may vary with the laser source.

Supercontinuum Generation with Oscillators

The relatively low pulse energy of a few nJ available directly from oscillators mandates confining light pulses to small cross-sections and incorporating guided wave geometries in order to enlarge the strength of the nonlinear interaction. While spectral broadening can already be achieved by utilizing SPM in standard telecommunication fibers [82], the amount of spectral broadening required for generation of octave spanning spectra can only be attained by focusing the light pulses to even smaller core sizes.

In this respect, the fabrication of a micro-structured fiber (MSF) with a ≈ 2 μm small core that is surrounded by air holes [83] has been a revolution. Indeed, such fibers led to the first observation of supercontinuum generation with oscillator pulse energies [56]. Almost simultaneously, supercontinuum generation has also been achieved in tapered standard telecommunication fibers, the radius of which has been reduced to ≈ 2 μm by pulling at the two sides of a piece of fiber while heating the region in between [84].

However, confining light to smaller cross-sections in order to enhance nonlinearity is not the only reason responsible for generation of such broadband spectra in these types of fibers. In addition, the zero dispersion wavelength of the fiber is shifted to the visible, resulting in anomalous dispersion[5] over the entire near infrared spectral region. In the presence of pure second-order anomalous dispersion, solitons may be formed. In the simplest case, these are wave packets of light that do not spread in time due to the counterbalancing action of dispersion and nonlinearity. As has

---

[5]GDD < 0, where GDD is the group delay dispersion.

been found by Husakou and Herrmann, soliton dynamics plays a decisive role for supercontinuum generation in the anomalous dispersion regime [85]. If the carrier frequency resides in the anomalous dispersion regime and the input pulse duration is less than ≈ 100 fs, supercontinuum generation can be decomposed into three steps (see, e.g., [86]). In the first stage, the pulses experience symmetric spectral broadening like in SPM which is accompanied by a strong temporal compression. Subsequently, a step called *soliton fission* takes place: the combination of input pulse power and fiber dispersion characteristics enables the input pulse to form a higher-order soliton. The presence of Raman scattering and higher-order dispersion, however, renders these higher-order solitons unstable such that fundamental first-order solitons are split off successively. Upon further propagation, these fundamental solitons experience a soliton self-frequency shift towards longer wavelengths as long wavelength spectral components are amplified by Raman scattering at the expense of short wavelength components [87, 88]. Simultaneous to the soliton self-frequency shift, the presence of higher-order dispersion leads to the generation of *dispersive waves* in the short wavelength, normal dispersion region due to phase matched four-wave mixing.

In total, these processes permit the generation of spectra that, e.g., cover more than one octave at the -10 dB level with 100 fs input pulses having 10 kW peak power (1 nJ pulse energy) [56]. Despite the strong nonlinear interaction experienced by the pulses, the CEP remains unchanged since all underlying nonlinearities are of four-wave mixing type. However, the strength of nonlinear interaction renders the CEP under certain circumstances sensitive to noise sources, which will be discussed in the Section 1.6.

Supercontinuum Generation with Amplified Pulses

The pulse energies of amplified Ti:sapphire pulses enable generation of supercontinua in bulk media, liquids or even gases. Most commonplace is the usage of bulk media with high bandgap energies such as sapphire or CaF_2. In these media the Kerr-effect leads to an initial self-focusing of the light beams, which enhances the nonlinear interaction. As a result, space-time focusing and self-steepening are encountered, which act to produce steep electric field gradients at the trailing edge of the pulse [89, 90]. Plasma formation by multiphoton ionization prevents an unlimited focusing of the beam due to the defocusing action of free electrons and sustains pulse propagation in a filamentary channel through the medium.

It has been reported that the amount of spectral broadening, especially to the blue, enhances the higher the bandgap of the medium is [91]. Furthermore, investigations showed a dependence on the numerical aperture of the focusing optics with optimum supercontinua obtained for low numerical apertures of ≈ 0.05 [92, 93]. Under such conditions, spectra extending from 450 nm to 1500 nm can be generated in bulk media with Ti:sapphire pulses having an energy from 50 nJ to 1 µJ and durations up to the 100 fs range [93].

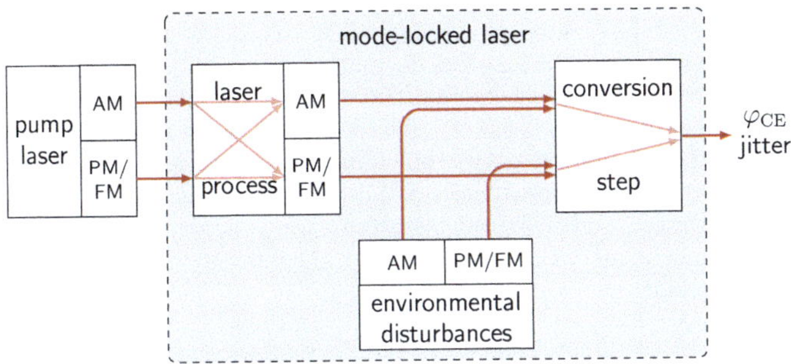

Figure 1.7.: Sources of φ_{CE} jitter in a mode-locked laser.

1.6. Origin of the Detected Carrier-Envelope Phase Jitter

The methods introduced in the last sections have enabled the successful stabilization of the CEP drift rate. However, several noise sources induce a residual CEP jitter. As a basis for the improvements discussed in the remainder of this thesis, the present section elucidates the impact of the individual noise sources in more detail.

Figure 1.7 depicts the generation of CEP jitter inside the mode-locked laser. In the laser process, amplitude modulation (AM) and phase modulation (PM) [or equivalently frequency modulation (FM)] of the pump laser are transfered into the AM and PM of the intracavity field. In general, the transfer process is not only due to a pure feed-through of the modulations [94], but also due to a conversion into each other. AM of the pump laser alters the spontaneous emission induced phase jitter. In turn, FM of the pump laser modulates the intracavity photon fluence $F_{\mathrm{ml}} \propto \alpha F_{\mathrm{pump}}$ of the mode-locked laser as a result of the frequency dependence of the absorption coefficient α and of the incident pump photon fluence F_{pump}. Typical pump lasers posses a fractional frequency instability and a relative intensity noise (RIN) on the 10^{-9} and 10^{-7} level, respectively. Hence, direct feed-through of the pump laser AM dominates the AM of the intracavity electric field.

Furthermore, environmental disturbances cause additional AM and PM of the intracavity field. For example, mirror vibrations convert to PM by resonator length variation and to AM by beam path misalignment. Variations of pressure and temperature lead to refractive index and dispersion changes.

The overall AM and PM of the intracavity field are converted to CEP jitter by three mechanisms. First, changes of the group-phase offset induced by refractive index changes including pump power dependent changes stemming from crystal heating [36]. Second, group-phase offset fluctuations by a combination of intracavity GDD and a spectral shift, which results from an adiabatic adaption of the solitary pulse to varied laser parameters [50, 95]. Third, changes of the nonlinear phase shift $\Delta\varphi_{\mathrm{nl}}$ that is imposed onto the pulse while forming the Kerr lens [36, 95, 96]. The latter can be further decomposed into a SPM term, which describes the phase shift $\Delta\varphi_{\mathrm{SPM}}$ of the carrier, and self-steepening term, which accounts for the temporal shift τ_{ss} of the maximum of the envelope due to the intensity dependent dispersion of the refractive index

$$\Delta\varphi_{\mathrm{nl}} = \Delta\varphi_{\mathrm{SPM}} + \omega_c\tau_{\mathrm{ss}} \,. \tag{1.15}$$

In the following, the pump power dependence of the per-roundtrip CEP shift $\Delta\varphi_{\mathrm{CE}} = \Delta\varphi_{\mathrm{GPO}} + \Delta\varphi_{\mathrm{nl}}$ is of special interest. Extending the formula derived by Sander et al. [96] by thermally induced group-phase offset variations, the sensitivity of CEP shift to variations of the pump power P can be written as[6]

$$\begin{aligned}
\frac{d\Delta\varphi_{\mathrm{CE}}}{dP} =\ & \left[\frac{\omega_c^2}{c}\int_0^L \left.\frac{dn(\omega,T,z)}{d\omega}\right|_{\omega=\omega_c} dz\right]\frac{dT}{dP} \\
& + \left[\int_0^L \left(\frac{2\omega_c}{c}\left.\frac{dn(\omega,T,z)}{d\omega}\right|_{\omega_c} + \frac{\omega_c^2}{c}\left.\frac{d^2n(\omega,T,z)}{d\omega^2}\right|_{\omega_c}\right)dz\right]\frac{d\omega_c}{dP} \\
& + \frac{d\Delta\varphi_{\mathrm{SPM}}}{dP} + \omega_c\frac{d\tau_{\mathrm{ss}}}{dP} + \tau_{\mathrm{ss}}\frac{d\omega_c}{dP} \,.
\end{aligned} \tag{1.16}$$

Here, the first term stems from the temperature induced variation of the group-phase offset, the second term describes the modification of the group-phase offset due to a carrier frequency shift, and the last three terms result from the power dependence of the phase shift resulting from nonlinear refraction and its dispersion.

The relative strength of these contributions has been subject of a number of studies. Xu et al. argued in Ref. [50] that the carrier frequency shift dependent variation of the per-roundtrip group-phase offset dominates over the SPM term (self-steepening was not considered). Indeed, in following measurements by Holman et al., a correlation between CEF changes and shifts of the carrier frequency laser has been observed in a Ti:sapphire [95]. As an analysis of the second term in Eq. (1.16) shows, the strength of the spectral shift dependence of the group-phase offset scales with the magnitude of the intracavity GDD [95], which is on the order of some 10 to 100 fs^2 in few-cycle oscillators [96]. In other words, the variation of the roundtrip time that is

---

[6]For the sake of clarity of the presentation, a spatially varying temperature change is not considered here. The description can easily be adapted to account for such changes by incorporating $dT(z)/dP$ into the first integral on the right-hand side of Eq. (1.16).

accompanied with a carrier frequency shift is the lesser the lower the variation of the group delay with carrier frequency is. Accordingly, Holman et al. reported that the CEF modulation amplitude decreases when the magnitude of net intracavity GDD is minimized. However, the minimization of the GDD also led to an increase of the oscillator's bandwidth such that the observed decrease in CEF modulation amplitude may also result from a complete filling of the gain bandwidth by the broader pulse spectrum. In the presence of such a complete filling, the gain bandwidth acts as a filter that limits the supported amount of spectral shift [72]. For the case of a dispersion managed, octave-spanning Ti:sapphire oscillator, Sander et al. showed by numerical simulations including Raman induced spectral shifts that the impact of the spectral shift term is an order of magnitude smaller than the effect of self-steepening. Furthermore, self-steepening was found to dominate over SPM in these oscillators such that the per-roundtrip CEP shift is almost exclusively determined by self-steepening [96]. Given the experimental results of Holman et al. [95] and the ≥ 100 nm bandwidth of the Ti:sapphire oscillators employed throughout this thesis, it is expected that nonlinear phase shift contributions dominate in the lasers used.

In addition to this true CEP jitter generated in the mode-locked laser, spurious noise is added in the detection process by supercontinuum generation and interferometer instability. In the supercontinuum generation step noise is amplified in the nonlinear interaction [80, 86, 97–99]. As a result of shot noise, quantum noise statistics of losses, and the input pulse noise, variations of the relative group delay (amplitude-to-phase coupling) or amplitude (excess AM) of the f and $2f$ components used for f–$2f$ interferometry are induced. This compromises the stability of the interference that is analyzed in f–$2f$ interferometers [100]. In a subsequent phase extraction step, spurious phase and amplitude modulation of the interference may then be interpreted as CEP noise, and, thus, the accuracy of the retrieved CEP is deteriorated. Similarly, path length fluctuations of the arms of the f–$2f$ interferometer due to mirror vibrations or dispersion changes lead to variations of the interference between f and $2f$ components. As components out of the wings of the spectrum are heterodyned, f–$2f$ interferometry is also confronted with low light levels. Hence, shot noise in the photodetection induces artificial amplitude modulations to the interference signal. In the case of, e.g., oscillator f–$2f$ interferometers, shot noise typically limits achievable signal-to-noise ratios to ≤ 50 dB in a 100 kHz bandwidth. In addition, these low light levels mandate usage of amplifying photodectetors, which involves excess noise, i.e., signal height fluctuations due to the variations of the amplification factor.

The presented discussion of the noise sources may suggests that any improvements of the respectable accuracy CEP stabilization setups achieved so far requires a revolution or even the manipulation of the light's quantum statistics. The following chapter, however, presents two refinements of the already existing technology, which are easy to implement, yet yield a considerable improvement.

Chapter 2

Improvements of the Carrier-Envelope Phase Stabilization of Amplifier Systems

Measurement and stabilization of the CEP drift of femtosecond pulse trains has found widespread application in frequency metrology [13] and high-field nonlinear optics. One of the major achievements of CEP control of femtosecond laser pulses is the generation of isolated attosecond laser pulses [9], for which the precise timing control of the field maximum within the envelope structure of few-cycle laser pulses is crucial. Even though these success stories were only achieved by continuously improving upon the degree of maturity of CEP stabilization over the last years, there is still room for improvement. Two of such improvements will be discussed in the following sections. These improvements remove the technical noise layer and give deeper insight into underlying physical noise sources.

In the first part of this chapter, two quasi-common-path interferometer topologies for detection of the CEP drift rate of oscillators are compared to the conventional two-path interferometer design in regard of their immunity towards external disturbances. The frequently used Mach-Zehnder interferometer is found to be easily corrupted by acoustic noise contributions and air streaks, whereas the novel quasi-common-path variants of the f–$2f$ interferometer exhibit a greatly enhanced immunity, leading to more than 40% reduction of residual noise. In the course of this thesis, these results will become particularly important for the light power variation detector which is devised in Chapter 4. As this power detector is based on the analysis of the CEP drift rate and reaches its highest sensitivity in the acoustic frequency range, prevention of spurious noise contribution, such as interferometer noise, is essential for achieving its extraordinary sensitivity.

In the second part of this chapter, a novel method for reducing residual CEP jitter of the slow loop stabilization of amplifiers is proposed. The key idea here is to directly extract the phase from spectral interferograms using a two-detector arrangement rather than a CCD. This method is demonstrated with single-shot measurement and stabilization of the CEP at the full 3 kHz repetition rate of the amplifier system. Furthermore, deeper insight into the physical origin of the residual phase jitter in amplifiers is obtained. It is found that, in the amplifier laser system used, fast CEP jitter is mainly inherited from the oscillator stabilization loop. Previously unreported indications for a rapid pulse-to-pulse jitter from the amplifier pump laser are also found.

The work presented in this chapter has partly been obtained in collaboration with Christian Grebing and with equal contributions of both of us. The results have partially been published in Refs. [101, 102] and in the thesis of Christian Grebing [103]. In the following, these results are extended and reevaluated in view of the novel material presented in Chapters 3 and 4.

2.1. Quasi-common-path f–$2f$ Interferometer

As already pointed out in the introductory chapter, measurement and stabilization of the CEP drift of femtosecond oscillators rely on heterodyning frequency-doubled long-wavelength components with fundamental components from the short-wavelength edge of an octave spanning frequency comb in an f–$2f$ interferometer [11]. The first demonstrations of the f–$2f$ interferometer utilized a Mach-Zehnder (MZ) type setup, see, e.g., Ref. [58]. This topology has probably found the most widespread application, in particular in commercial devices. As the MZ geometry is difficult to miniaturize and as its rather large arm lengths are quite susceptible to acoustic noise, air streaks and thermal drift, some authors have resorted to a Michelson-type geometry [104]. However, it appears difficult to reduce the arm lengths to below a few centimeters for any kind of two-path interferometer. Moreover, it was recently reported that spurious interferometer noise may contribute to nearly half of the residual phase drift of a stabilized amplifier system [105]. This finding explains to some extent why in-loop (IL) measurements often appear to indicate 10 times smaller CEP noise values than out-of-loop (OOL) characterizations [69]. Consequently, a more stable setup of the interferometer appears crucial for achieving better CEP stabilization.

The issue of interferometer noise is easily avoided by an inline arrangement or common-path interferometer. While this concept is commonplace for measurement of the CEP for amplified kHz-repetition rate laser systems [53], common-path arrangements for oscillator CEP stabilization are rare. The major obstacle for using an inline geometry is the necessity for vanishing group delay between the f and the $2f$ component. For using heterodyne detection, the group delay difference either has to be on the order of the pulse width [75], or it needs to be compensated. In particular, when an MSF is used for spectral broadening, the group delay between f

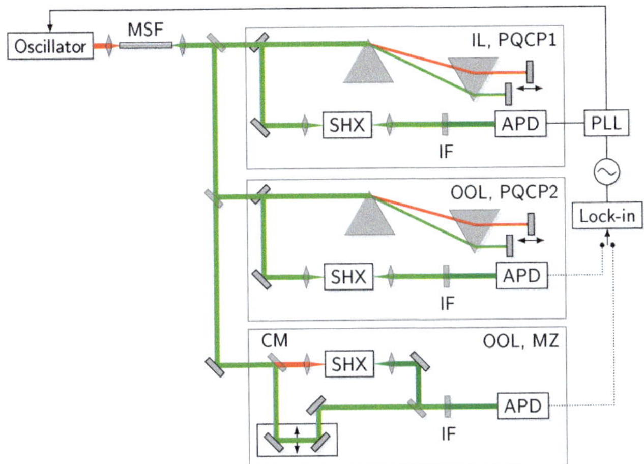

Figure 2.1.: Illustration of the measurement setup. MSF: micro-structured fiber for spectral broadening. SHX: second harmonic crystal for 1060 nm. APD: avalanche photodiode for detecting the beat signal. PLL: phase locked loop. CM: cold mirror. The prisms are separated by about 17 cm in order to achieve the required spatial separation. Furthermore, splitting of the incoming and retro-reflected beam is obtained by a bisected end mirror that is slightly tilted vertically.

and $2f$ components easily amounts to picoseconds, requiring several thousand fs^2 for its compensation. Thus, true common-path interferometers were only applied utilizing octave-spanning pulses [74, 106], in a monolithic f–0 setup [54, 73], or in fiber lasers by compensating the group delay accumulated during spectral broadening in a piece of fiber with inverse group delay dispersion [107, 108]. While the former two implementations required achieving an intricate compromise between the oscillators bandwidth and its stability of operation, which renders it only applicable to a limited class of lasers, the latter is rather static and inflexible such that it cannot cope with dynamic group delay variations in the supercontinuum generation step (see Section 1.6).

In the following, the performance of two quasi-common-path (QCP) interferometer layouts is discussed, and it is shown that these interferometer configurations offer a convenient way for compensation of the group delay difference while maintaining the noise immunity of true inline interferometers and the versatility of two-path interferometers.

For a systematic comparison of the noise sensitivity of different interferometers, an OOL measurement scheme is employed [see, e.g., Fig. 2.1]. For obtaining octave spectral coverage, the output of a commercial Ti:sapphire oscillator with 10 fs pulse duration is broadened in a 20 cm long MSF. Then, the interferometer phase noise is characterized using two interferometers: one for generating a phase error signal for CEP drift rate stabilization (IL interferometer) and a second one for an independent check of the CEP drift rate (OOL interferometer). The CEP drift of the OOL interferometer is retrieved by means of a radio-frequency lock-in measurement. Possible CEP noise contributions from the MSF are deliberately eliminated by deriving the two beams for the IL and OOL interferometer after the MSF. This ensures an exclusive comparison of interferometer drifts of the different topologies.

2.1.1. Prism-based Quasi-Common-Path f–$2f$ Interferometer

In the prism-based quasi-common-path (PQCP) interferometer investigated here [PQCP1 and PQCP2 in Fig. 2.1], the spectral broadening induced group delay of about 1.5 ps between fundamental and harmonics is compensated making use of a spatial separation of the spectral components by means of a heavy-flint Brewster prism sequence (see PQCP1 in Fig. 2.1) [109, 110]. In the experimental implementation the prism's apex separation amounts to 17 cm and the prism sequence is arranged to separate infrared portions at 1060 nm and portions of the continuum at the second harmonic wavelength of 530 nm. The infrared and visible light is then separately retro-reflected under a small vertical angle by a bisected end mirror. A path difference of 4 mm is needed to compensate for the group delay differences generated in the MSF. The total unshared, but parallely propagated path between prism and retro-reflector amounts to $\approx$ 55 mm. The bisected mirror is the only optical element that both arms do not share commonly. The retro-reflected beam is focused by an $f = 18$ mm lens into a periodically poled lithium niobate (PPLN), optimized for efficient 1060 nm frequency doubling. The generated 530 nm light is spectrally filtered and detected by an APD.

Since the PQCP interferometer is anticipated to show a better noise performance, the PQCP interferometer PQCP1 is used as IL interferometer for stabilizing the CEP drift rate. Then the OOL measurement is conducted with either a second PQCP interferometer (PQCP2) or an MZ interferometer as OOL interferometer. In each of these interferometers a comparable beat note signal level of $\approx$ 30 dB above noise floor [resolution bandwidth (RBW) 100 kHz, video bandwidth (VBW) 300 kHz] is obtained.

For comparison, a conventional MZ type interferometer is also set up (see Fig. 2.1). Here fundamental and harmonics are split into the two interferometer arms by a dichroic mirror. Subsequently, frequency doubling as well as adjustment of the temporal delay are easily realized in separate arms. As already pointed out, the distinct spatial separation of both arms and the great number of optical elements not shared commonly renders the MZ setup sensitive to external perturbations, e.g., air streaks and vibrations caused by pumps and water chillers in the lab. Nevertheless, the length of the separated beam paths was kept to a minimum value of about 20 cm.

As a first test, the enclosures of the OOL interferometers have been removed in order to operate the interferometers without shielding against acoustics and air movements. This test has revealed the considerably better performance of the PQCP geometry by the suppression of pronounced drifts and strong phase excursion experienced by the MZ interferometer.

In a second experiment, external disturbances are eliminated by enclosing the OOL interferometer, and the measurements are redone. The time series of the phase noise measured OOL with both, the MZ and PQCP setup are shown in Fig. 2.2(a). Additionally, the error voltage of the PLL is recorded for an evaluation of the IL CEP noise. As already reported by others [69], the IL standard deviation underestimates the OOL values by almost one order of magnitude. This discrepancy corroborates the necessity of OOL measurements for a reliable evaluation of the residual CEP drift.

A comparison of the OOL time series reveals the better performance of the PQCP geometry, which immediately results in $> 40\%$ reduced residual CEP noise of 150 mrad compared to the MZ setup. In the spectral analysis of the phase noise [Fig. 2.2(b)], mainly a noise reduction in the acoustic band above 100 Hz is observed. This difference becomes clearer when looking at the integrated phase noise depicted in Fig. 2.2(c). In this representation, the phase noise density is integrated from the sampling rate to the inverse observation time $1/\Delta t$, such that contributions of the peaks in MZ trace in Fig. 2.2(b) show up as pronounced steps at about 70, 200, 300, and 450 Hz. As these steps appear only in the MZ trace at a step height of about 10 mrad, they are attributed to forced mirror vibrations caused by acoustic noise in the lab, which affect the MZ setup more strongly than the PQCP. It is readily calculated that, in the wavelength region of interest, a step height of about 10 mrad corresponds to tiny mirror displacements of less than 1 nm root mean square (RMS). Alternatively, the differing noise susceptibility of the two interferometer topologies may also be explained by air streaks in the interferometers, i.e., induced small beam deflections resulting from thermally or acoustically caused refractive index variations across the beam paths. Given the shielding of the interferometers, however, the impact of air streaks is expected to be negligible in this measurement.

Apart from the acoustically induced distortions, the strongest step-like feature in Fig. 2.2(b), however, appears at an identical height of about 110 mrad at 100 Hz in both curves. The particular frequency and its appearance in both interferometers makes an acoustic origin very unlikely. The likely origin of this noise contribution could be traced back to residual power supply ripple of about 100 $\mu\mathrm{V}/\sqrt{\mathrm{Hz}}$ at 100 Hz in the commercial locking electronics used.

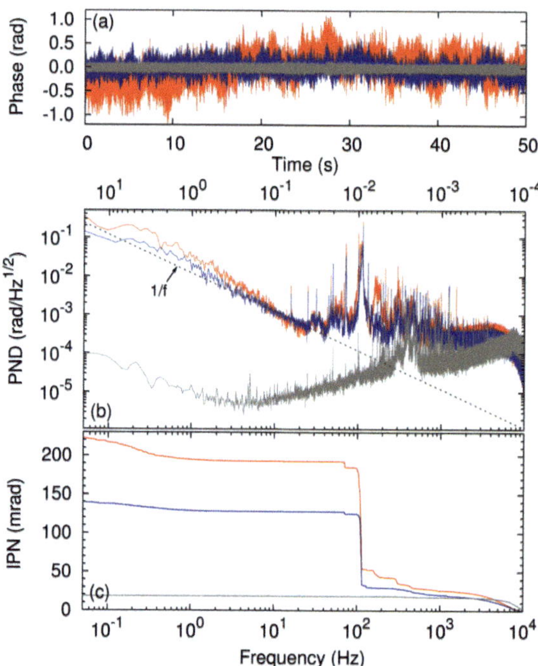

Figure 2.2.: Measurement results in shielded mode. Blue: PQCP; red: MZ; gray: in-loop. (a) 50 s segment of a 120 s long phase error time series. Residual RMS phase noise over the entire time series: 240 mrad (MZ), 150 mrad (QCP), 20 mrad (in-loop). (b) Phase noise density (PND) vs. frequency. The black dotted curve shows that the low frequency data are well described by an $1/f$-like noise contribution. (c) Integrated phase noise (IPN) vs. observation time (cf. appendix A.1 for a definition).

Finally, the integrated phase noise plots also show a slow increase for long observation times, which is again slightly more pronounced for the MZ interferometer than for the PQCP setup. By extrapolating the course of the phase noise density (PND) for frequencies below 20 Hz by the $1/f$-fit shown in Fig. 2.2, the accumulated CEP error is estimated to exceed 2π after an integration time of 77 hours.

The joint experiments with Christian Grebing revealed that a PQCP does in fact perform better than a simple MZ. Yet, the improvement is not quite as dramatic as originally expected; in particular when looking at the slow drift-like changes which manifest themselves in the $1/f$ rolloff in the noise spectra. As the PQCP is practically immune against temperature or pressure-induced drifts of the refractive index, the question on the origin of the residual $1/f$ characteristics arises. One possible origin is beam-pointing instabilities of the beam entering the interferometer. As the PQCP contains elements with angular dispersion, beam pointing variations may translate into CEP variations. Such coupling mechanisms have already been found to be responsible for the inferior phase noise performance of prism-based oscillators [36]. Given the coupling values determined for prism-compressors based oscillators, an order of magnitude estimation yields that even tiny angle deviations on the order of 1 μrad can readily induce phase errors of ~ 10 mrad. To check on this possibility, a novel beam-pointing immune QCP geometry is investigated in the next subsection. These experiments will also avoid the observed 100 Hz artifact stemming from the digital locking electronics. The practical absence of this artifact when using an analog locking electronics further underlines the finding that the corresponding feature in the PNDs originates from the power supply ripple of the digital electronics.

2.1.2. Dichroic Quasi-Common-Path $f{-}2f$ Interferometer

The alternative QCP interferometer is shown in the top box of Fig. 2.3 and replaces the prism-based wavelength separation by the use of a dichroic mirror. Therefore, it is termed dichroic quasi-common-path (DQCP). The dichroic mirror reflects the green-wing part of the spectrum and is followed by a second mirror reflecting the transmitted infrared light. By variation of the mirror's spatial separation, the relative timing between the f and $2f$ component used for $f{-}2f$-interferometry is adjusted. In this configuration the angular dispersion of the spectral components present in the PQCP interferometer is removed and the unshared path is further reduced to the ≈ 2 mm offset of the end mirrors.

In the experimental implementation, the mirrors are mounted on optical rails for maximum stability. Furthermore, an analog PLL is employed for CEP stabilization in order to avoid spurious CEP noise contributions stemming from the power supply ripple of the commercial, digital-PLL-based locking electronics. Despite inducing power supply ripples, the increased hold-in range of 128π of the digital PLL is associated with a higher phase uncertainty. Due to the larger phase range, unavoidable ripples on the servo voltage translate into a ≈ 120 times higher CEP error

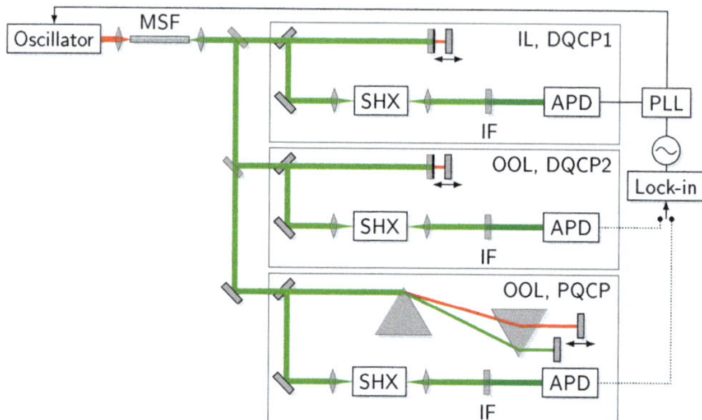

Figure 2.3.: Out-of-loop measurement scheme used for the performance characterization of the dichroic quasi-common-path interferometer (DQCP). Both DQCP 1 and DQCP 2 are build with the same optical mounts. The end mirrors are mounted on optical rails for ensuring the lowest relative movement possible in an non-monolithic setup. The spatial displacement between the end mirrors amounts to $1 - 2$ mm.

than in the analog PLL, which only allows for an unambiguous phase retrieval in an interval with a span of π. Therefore, the analog PLL also permits a considerably tighter lock making the evaluation of interferometer noise more accurate. As in the previous performance comparison of the PQCP and MZ interferometer, an OOL scheme is utilized to characterize the performance of the DQCP interferometer. Here, the DQCP configuration is used as the IL interferometer. A beat note signal-to-noise ratio of ≈ 40 dB (RBW 100 kHz, VBW 300 kHz) is obtained in all three interferometers. Running the DQCP interferometer unshielded showed that it is as immune to air streaks as the PQCP interferometer.

The OOL phase error characterization measurements in shielded and stabilized operation mode are plotted in Fig. 2.4. Already the time series in Fig. 2.4(a) reveals the considerably superior stabilization performance obtained with the DQCP configuration. While the PQCP trace exhibits excursion with maximum deviations up to 350 mrad, the DQCP trace mainly consists of a continuous noise band with just slight deviations from it. The corresponding PND is plotted in Fig. 2.4(b) and clarifies that this discrepancy chiefly results from the low-frequency features, as for the high frequencies above ≈ 100 Hz no pronounced difference is visible. While the PQCP curve starts to deviate from the high-frequency plateau for frequencies below ≈ 50 Hz, this deviation starts to set in around 10 Hz in the DQCP measurement.

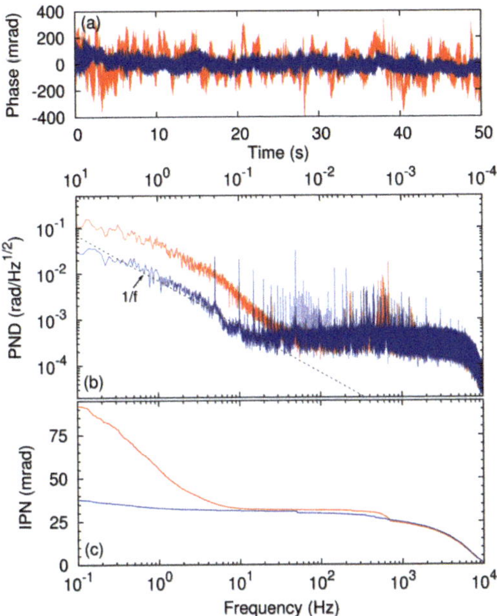

Figure 2.4.: Out-of-loop performance characterization of the dichroic quasi-common-path interferometer. (a) 50 s segment of a 120 s long phase error time series. Blue: dichroic quasi-common-path interferometer; red: quasi-common-path interferometer. (b) Phase noise densities (PND). (c) Integrated phase noise (IPN).

Towards lower frequencies the DQCP curve follows the PQCP curve with a factor of ≈ 4 reduced PND. This reduction manifests itself in a considerably less pronounced increase of the integrated phase noise (IPN) towards lower frequencies [Fig. 2.4(c)]. While the PQCP interferometer exhibits a phase error increase of ≈ 55 mrad from 10 Hz to 0.1 Hz, the OOL characterization only reveals an IPN increase of < 10 mrad for the DQCP configuration in this frequency range. For the case of the DQCP interferometer, extrapolation of the PND for frequencies below 0.3 Hz by the $1/f$-fit shown in Fig. 2.4 yields that the time span until surpassing a phase error of 2π is extended to 230 hours.

The likely origin of the superior long-term performance of the DQCP configuration is the removal of the angular dispersion of the beam. Besides making the interferometer even more common-path, this allows for mounting of the end mir-

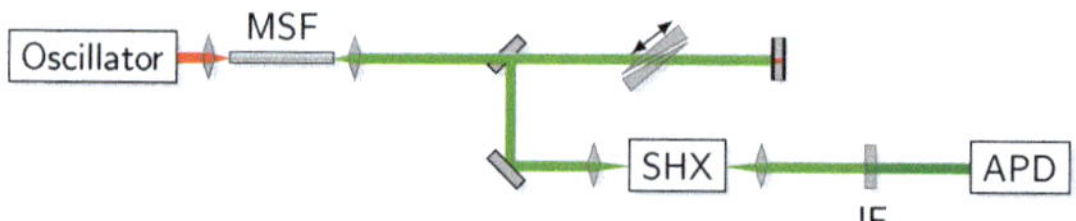

Figure 2.5.: Setup for a monolithic dichroic quasi-common-path interferometer, where the dichroic end mirrors with variable spatial separation are replaced by a monolithic end mirror combination. MSF: microstructured fiber; SHX: second harmonic crystal; IF: interference filter; APD: avalanche photodiode.

rors on the same foundation, such that the relative drift is lessened. Furthermore, beam-pointing-to-CEP coupling in the prism sequence [36] of the PQCP is removed in the DQCP configuration. Probably the combination of both effects leads to the observed improvement.

2.1.3. Further Improvements

In order to further improve its noise properties, the layout of the DQCP interferometer can be rendered monolithic as shown in Fig. 2.5. In this layout an end mirror whose front and back surface are coated for reflecting the green and the red, respectively, is used to induce a fixed delay between the f and $2f$ spectral component by the additional optical path for the red wing part of the spectrum. If this is made to overcompensate the temporal delay induced by the MSF, variable material dispersion in a pair of wedges can be used to obtain zero delay again. Compared to the DQCP interferometer, the monolithic design will render the interference free of relative end mirror drift, and, as the separation of the wedges is small, beam pointing induced phase errors are negligible and will not deteriorate the performance. Such an layout is as simple as the birefringence based common-path interferometer topologies [111, 112] that have been developed simultaneous to this study and allows utilizing quasi-phase matching for second harmonic generation.

Depending on the characteristics of the MSF and the input pulses, however, the interferometer may be drastically simplified further by choosing a different wavelength combination for heterodyning. For example for the MSF used in the experiment, a closer look at the fiber induced delay [Fig. 2.6(a)] reveals that, for a fundamental wavelength $\lambda \approx 580$ nm, there exists an f–$2f$ combination that does not exhibit a delay between the f and $2f$ component. However, the experimental implementation of such an exact delay matching requires consideration of the dispersion introduced by the elements following the fiber (lenses, filter, frequency doubling crystal) since the zero delay has to be achieved on the photodetector. Therefore, it seems beneficial to choose a wavelength combination for which the fundamental wavelength is closer

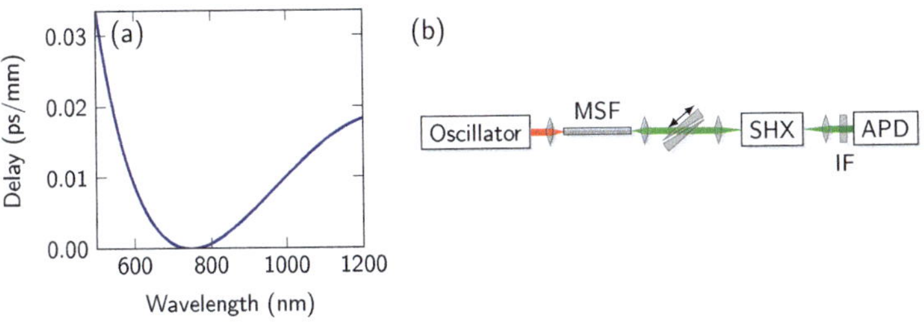

Figure 2.6.: Delay of spectral components with respect to the zero-dispersion wavelength induced by the micro-structured fiber (NL-PM 750, NKT Photonics) used in the experiments. MSF: micro-structured fiber; SHX: second harmonic crystal; IF: interference filter; APD: avalanche photodiode.

to the zero dispersion wavelength. For such an f–$2f$ combination, the long wavelength components to be frequency doubled are delayed with respect to the short wavelength components, as the delay accumulated by the long wavelength components monotonically increase up to a wavelength > 1200 nm. Therefore, adjustment of the delay of the interfering pulses on the photodetector is easily achievable employing variable material dispersion in, e.g., a pair of wedges. The generation of supercontinua with significant contributions for such long wavelength has already been achieved, see, e.g., the spectrum shown in Fig. 1(b) of Ref. [111]. Thus, such an f–$2f$ combination in combination with the simple interferometer configuration shown in Fig. 2.6(b) is ready to be tested in future.

2.2. Fast f–$2f$ Interferometer for Amplifier Stabilization

In the previous section, revised interferometer designs enabled the reduction of the remaining CEP jitter in oscillator CEP stabilization. Applications such as attosecond pulses generation [9], however, require CEP stabilization of femtosecond amplified laser pulses with mJ energy. During amplification, the pulse train's CEP stability may be compromised by beam pointing variations in stretcher and compressor [36, 37], pulse picking [76], and the amplification itself [80]. Thus, further progress in amplifier CEP stabilization demands pinpointing whether one of these noise generating mechanism dominates or whether the residual CEP jitter is inherited from the seed oscillator. However, discerning these mechanisms requires resolving the pulse-to-pulse CEP jitter at a few kHz repetition rate, which is not viable with currently available phase extraction setups.

So far, the CEP of amplified laser pulses has been measured by decoding the fringe position of the f–$2f$ spectral interference pattern [53, 59]. To this end, the interference pattern is read out by CCD cameras or diode arrays mounted in the image plane of a spectrometer and post-processed with a computer. Even though line-scan cameras exist that can read out such fringe patterns with kHz acquisition rates, low light levels typically mandate averaging over at least 10 laser shots. Otherwise the poor signal-to-shot-noise ratio prohibits meaningful data extraction. With multi-shot averaged RMS jitters of CEP stabilized systems lying in the $100 - 300$ mrad range [80, 81], pulse-to-pulse fluctuations of the CEP may therefore readily amount to 1 rad, provided that the underlying noise generating mechanism obeys Gaussian statistics. In fact, recent single-shot measurements with the standard setup reported CEP jitters well above one radian [113], which is an indication for the dominance of shot noise. These findings raise the question whether a meaningful single-shot CEP detection is feasible after all.

In the following, a CEP extraction principle is proposed and demonstrated that utilizes a two-detector arrangement. Compared to the detection of the interference pattern by a CCD, where each pixel acts as an individual detector, such a two-detector arrangement enables achieving an enhanced signal-to-shot-noise ratio. Furthermore, the setup completely relies on analog electronics and obviates the need for numerical post-processing for CEP retrieval. Hence, delays that are caused by the numerical post-processing step are eliminated. As a result, the limited phase update rates of some 10 Hz in real-time applications are overcome, and single-shot CEP measurement and stabilization at kHz update rates is achieved for the first time.

Scheme for Shot-Noise Limited Carrier-Envelope Phase Detection

The optical setup of the proposed f–$2f$ interferometer shown in Fig. 2.7 widely resembles the original common-path interferometer proposed by Kakehata et al. [53]. For this experiment, a commercially available 3 kHz Ti:sapphire amplifier system is utilized, which is pumped by a diode-pumped q-switched Nd:YAG laser. The CPA amplifier exhibits residual pulse energy fluctuations of $< 1.8\%$ and is seeded by an oscillator that is CEP stabilized employing a MZ interferometer. The power of the oscillator pump laser is modulated in order to stabilize the CEP of both, oscillator and amplifier. For measuring the CEP phase of the amplifier, a fraction of its output is spectrally broadened by supercontinuum generation in a sapphire plate. Using a variable attenuator as well as a variable aperture in front of the focusing optics the spectrum is optimized for f–$2f$ interferometry [92]. The supercontinuum generated in the sapphire plate is focused into a 1 cm long LBO crystal, cut for non-critically phase-matched second harmonic generation (SHG) at 1000 nm. Non-critical phase matching was chosen to avoid walk-off effects. The phase-matching supports about

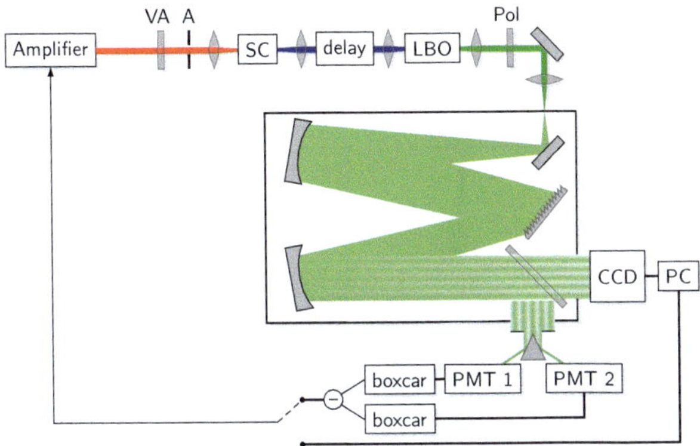

Figure 2.7.: Illustration of the measurement setup. VA: variable attenuator. A: aperture. SC: supercontinuum generation in a sapphire plate. delay: dispersive delay introduced in a piece of fused silica. LBO: frequency-doubling in lithium triborate crystal, phase-matched at 1060 nm. Pol: polarizer.

6 nm bandwidth. With the orthogonally polarized second harmonic, a polarizer is used for projection of both 500 nm components on the same axis. The resulting spectral interference is then analyzed in a half-meter spectrograph. A dispersive glass block was inserted between supercontinuum generation and LBO crystal for adjusting the fringe spacing to approximately 1 nm, which is compatible with fringe pattern analysis using both, the analog scheme and a CCD camera.

With the insertion of an inconel beam splitter, the spectrograph is equipped with a second output port allowing simultaneous detection with an analog and a digital scheme. While in one port a kHz-rate line-scan CCD camera is placed, an adjustable slit assembly with calibrated micrometer set screw is mounted in the second port. Furthermore, a miniature metal prism is introduced directly behind the slit assembly, such that the apex of the prism comes to lie symmetrically in between the blades of the slit assembly. The two polished faces of the prism direct the light onto two identical bialkali photomultiplier tubes, exhibiting a 30% quantum efficiency in the wavelength region of interest. Signals from the multiplier are read out by two gated integrators and finally subtracted from each other. By disengaging the oscillator stabilization circuit, the prism position is first adjusted for symmetric beam splitting onto the two photomultipliers.

As illustrated in Fig. 2.8(a) and (b), this setup allows for a retrieval of φ_{CE}, since the changing phasing of the fringe pattern with respect to the center of the

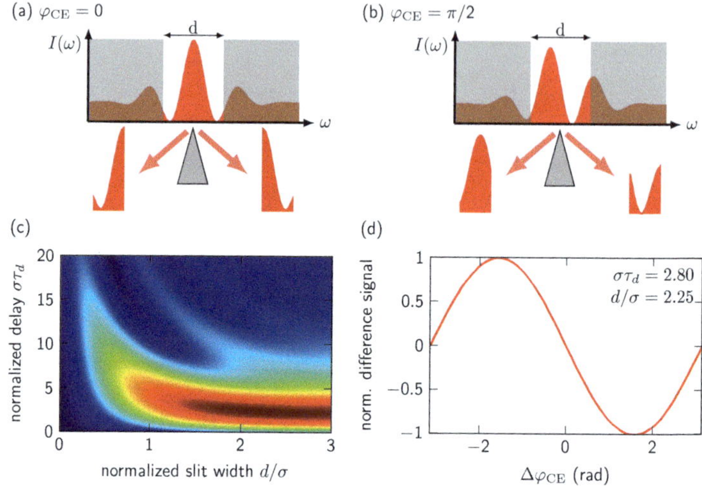

Figure 2.8.: Principle of the analog carrier-envelope phase retrieval. The case of symmetric and asymmetric splitting of the interference pattern is shown in (a) and (b), respectively. (c) Variation of the φ_{CE}-contrast in false colors for different slit widths and delays under the assumption that the second harmonic spectrum has a Gaussian shape $I_{SH} \propto \exp\left(2(\omega - \omega_0)^2/\sigma^2\right)$ and the spectrum of the fundamental component is constant $I_{fund} \approx$ const. Here, red indicates a high contrast and blue a low contrast. (d) Sinusoidal variation of the detected difference signal when varying φ_{CE}.

slit assembly leads to different amount of light detected by the two photomultiplier tubes. An optimum contrast for changing φ_{CE} is achieved when the total detection window width is on the order of the second harmonic phase matching bandwidth and when the fringe pattern period is set to a similar value [see Fig. 2.8(c)]. The difference voltage varies sinusoidally with the CEP encoded in the fringe pattern, which allows for unambiguous determination of the CEP in a $[-\pi/2, \pi/2]$ interval [Fig. 2.8(d)].

Experimental Results

Figure 2.9 shows a comparison of the signals obtained from the digital (5 ms integration time, $\approx$ 40 Hz update rate) and the analog detection scheme, with only

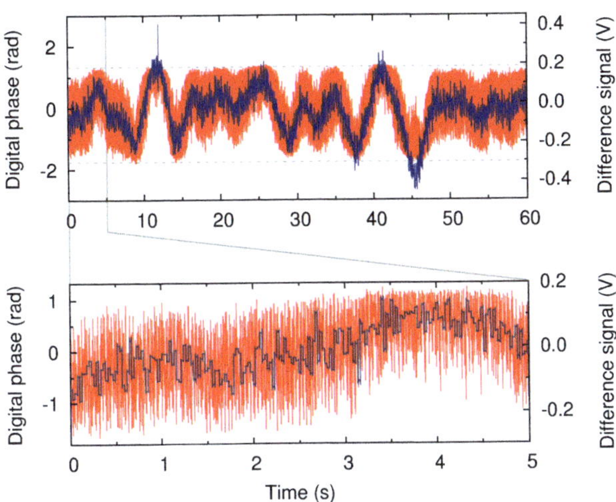

Figure 2.9.: Comparison of the signal obtained with the digital setup (blue) to the signal obtained by the novel analog setup (red) while amplifier stabilization has been switched off. The two vertical gray lines mark the interval where a unambiguous inversion is possible. In the following, this is utilized to convert the measured voltages into phases.

the oscillator servo loop locked. Both signals are strongly correlated, which corroborates the validity of the analog detection scheme. In the enlarged section shown in Fig. 2.9(b), fast noise contributions are observable in the analog signal. As will be detailed below, these contributions are clearly detectable well above the shot-noise level of the setup. However, they are completely hidden to the digital setup.

In a second step, the analog signal is directly utilized for CEP stabilization of the amplifier. To this end, a servo voltage is generated by processing the analog difference signal with a commercial PID controller. The output of the gated integrators is only filtered with a 3 kHz low-pass filter to remove reset glitches of the boxcar averager. The achieved stabilization performance is illustrated in Fig. 2.10(a), with an RMS error of the digital signal of 115 mrad over 10 min integration time, which is close to the best reported values in literature [80, 81]. However, pulse-to-pulse analysis with the analog setup reveals 210 mrad residual phase jitter, with a non-Gaussian distribution of the servo error. Despite the excellent performance, occasional single-shot excursions of the CEP phase amount to 0.5 mrad, which is completely hidden to multishot averaged CEP detection schemes.

For a deeper understanding of the underlying residual noise mechanisms, the data in Fig. 2.10(a) is Fourier analyzed. The phase noise density of the analog signal is

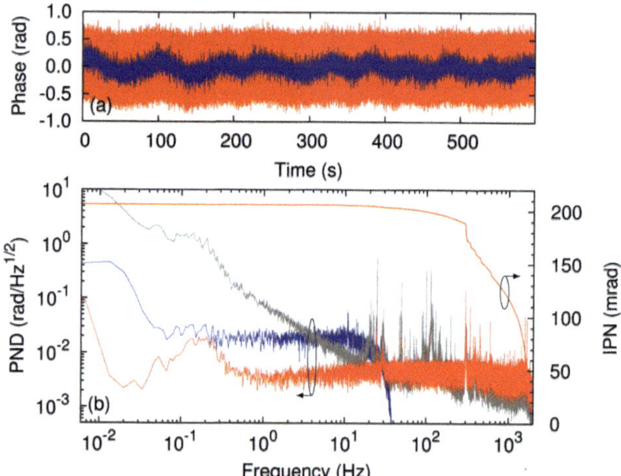

Figure 2.10.: (a) Comparison of the digital (blue) and the calibrated analog (red) signal after stabilization to the analog signal has been switched on. (b) One-sided Welch's phase noise density (PND) (cf. appendix A.1) of the analog signal (red), the digital signal (red) and of the oscillator measured out-of-loop (gray). Additionally, the integrated phase noise (IPN) of the analog signal is shown (orange, for a definition refer to appendix A.1). Note, that this analysis also reveals small steps in $200 - 600$ Hz frequency region, as observed in the comparison between quasi-common path and Mach-Zehnder interferometer topologies presented in the last section. These steps probably stem from the fact that these measurements were acquired while the oscillator was stabilized to a Mach-Zehnder interferometer.

presented in Fig. 2.10(b) and shows a wide plateau. On top of this plateau a broad peak is observed in the 0.2 Hz region, which is attributed to the periodic opening of the water valves in the cooling circuits. Furthermore, several sharp lines appear above 20 Hz that are normally hidden to the standard measurement. To most part, these noise lines also appear in the oscillator CEP noise spectrum. This indicates that the residual CEP noise of the amplifier is mainly limited by the performance of the oscillator stabilization, with only little influence from the amplifier pump laser. The integrated phase noise of the analog signal reveals that virtually all phase noise is accumulated within 5 ms, i.e., a time span that is equal to the integration of the CCD camera. This hidden rapid noise contribution also explains the discrepancy between the RMS phase errors measured with both methods and corroborates that a pulse-to-pulse analysis is necessary to reliably detect the CEP drift.

As it widely avoids technical detection noise contributions, such as read-out and thermal noise of CCD cameras, the photomultiplier-based approach is certainly more sensitive than a fast CCD camera. Yet, it should be pointed out that any scheme is ultimately limited by the total number of photons N detected in a single shot. Shot noise limits the detectable phase difference between two shots in the two-detector arrangement to $\Delta\varphi_{\mathrm{CE}} = \pi/\sqrt{N}$, which translates into an upper estimate $\Delta\varphi_{\mathrm{CE}} \leq 20$ mrad, given the signal levels in this experiment ($N \approx 32000$). Placing a red filter behind the sapphire plate and compensating for the signal decrease by tilting the polarizer allows for an evaluation of the total noise of both, electronic and optical origin. By doing so, it is found that in the analog detection scheme the spurious total CEP noise corresponds to approximately 40 mrad. Furthermore, the read-out noise introduced spurious CEP noise of the digital detection scheme already amounts to 65 mrad for an integration time of 5 ms. This noise is measured by blocking the beam entering the spectrometer. Given that the bandwidth of the analog detector is 100 times higher, the phase noise background of the CCD camera is therefore about 10 times higher when identical integration times are considered.

Finally, let us compare the fast f–$2f$ detection scheme to the CEP measurement by detection of the left-right asymmetry of ATI of atoms in the gas phase [60], which has recently been developed further to a single-shot technique [65, 114]. To the end of facilitating the CEP retrieval, a special method analyzing the left-right asymmetry in the plateau region of the ATI spectra has been developed [65]. For this analysis, the high energy plateau region is divided into two energy intervals and the number electrons detected on the left ($N_L^{(i)}$) and the right ($N_R^{(i)}$) detector in the energy interval $i = 1, 2$ is determined. Then, the asymmetry parameters

$$x^{(i)} = (N_L^{(i)} - N_R^{(i)})/(N_L^{(i)} + N_R^{(i)}) \tag{2.1}$$

are calculated. Thereafter, the CEP is retrieved via applying the same mapping to the results of numerical simulations [65]. The interval bounds can be chosen such that the parametrized coordinates $x^{(i)}$ vary sinusoidally with the CEP exhibiting a phase difference of $\pi/2$ [65]. For such a choice of mapping, stereo ATI measurements of pulses with varied CEP lie on a circle in the $x^{(1)}$-$x^{(2)}$ parameter space such that the CEP φ_{CE} depends linearly on the polar angle [114]

$$\varphi_{\mathrm{CE}} = \varphi_{\mathrm{offset}} + \arctan\left(x^{(1)}/x^{(2)}\right) , \tag{2.2}$$

where $\varphi_{\mathrm{offset}}$ is retrieved by comparison with numerical simulations. While the details of the procedure remain unspecified throughout the paper, Sayler et al. deduced an uncertainty of $\approx 115\,\mathrm{mrad}$ for an optimal choice of the interval bounds from averaged spectra collected with a non-CEP stabilized laser by means of Monte Carlo simulations [114]. At first sight, this seems to be a quite optimistic estimate of the uncertainty of the measurement, given that the uncertainty resulting from shot noise for the fast f–$2f$ detection scheme is 20 mrad in a single shot and the expected count rates of ATI electrons are considerably lower than that of photons in the f–$2f$ interferometer. To the end of understanding this puzzling finding, one can derive an estimate for the shot-noise-induced uncertainty in the stereo ATI technique. For the optimized choice of interval bounds, the uncertainty is roughly equally distributed along the polar coordinate in $x^{(1)}$-$x^{(2)}$ parameter space. Thus, one can choose an arbitrary point on the circle that the measurement values describe for estimating the uncertainty, e.g., the point $x^{(2)} = 0$ and $x^{(1)} = r$ (r is the radius of the circle). For this point the phase uncertainty $\Delta\varphi_{\mathrm{CE}}$ is mainly induced by variations in $x^{(2)}$ and since for $x^{(2)} = 0$ $N_R^{(2)} = N_L^{(2)}$, the shot-noise-induced uncertainty $\Delta\varphi_{\mathrm{CE}}$ is

$$\Delta\varphi_{\mathrm{CE}} \approx \frac{1}{r\sqrt{N_{\mathrm{el}}}}, \tag{2.3}$$

where $N_{\mathrm{el}} = N_R^{(2)}$. Thus, for the experimental conditions with $r \approx 0.75$ already $N_{\mathrm{el}} = 144$ electrons in either of the integration intervals is sufficient to explain a phase uncertainty of 115 mrad. Given the fact that earlier stereo ATI measurements only detected 50 electrons per shot [60], the low phase uncertainty achieved in the most recent experiments seems to be already stemming from a considerable improvement of the setup, such that achieving even higher electron count rates are only accessible using even shorter and more powerful laser pulses. Hence, a further decrease of the phase uncertainty in the stereo ATI setup seems to be limited, such that the fast f–$2f$ detection technique currently seems to be the only viable method for achieving a low-uncertainty, single-shot CEP retrieval of amplified laser pulses, especially for lasers that do not deliver ultrashort high-power pulses.

Further improvements

Further improvement of the detectivity of fast f–$2f$ detection scheme ultimately requires the increase of photon numbers in the detection. The supercontinuum process is the major bottleneck, since increasing power levels incident on the sapphire plate results in multi-filamentation destroying the CEP information [80] or even damage. Replacing the sapphire plate by materials with larger bandgap like CaF_2 or MgF_2 [91] or employing YAG [93] may enable enhancement of power levels. Also directly using the supercontinuum from a hollow fiber appears as a promising alternative, since a shortening of the amplified pulse duration is typically needed in the study of CEP dependent phenomena, anyway.

However, usage of the demonstrated detection system behind a hollow fiber compressor is not optimal for two reasons. First, the setup requires a vanishing delay between the heterodyned components in order to achieve a large spatial fringe period for optimum detectivity. Thus, a fraction of the precious high-power few-cycle pulse has to be tapped off for CEP detection. More elegant would be the usage of the uncompressed leakage through the chirped mirrors, which are typically used for recompressing the spectrally broadened output from the hollow fiber. Since an additional non-common-path layout of delay compensation inside the f–$2f$ interferometer enhances interferometer noise and compensation of the delay of spectrally far apart components utilizing chirped mirrors is accompanied with strong losses, adaption of the detection to enable CEP retrieval with non-vanishing group delay seems to be the only viable way for single-shot CEP detection in the leaking beam. Second, the narrowness of the spectral region analyzed by the single-shot CEP retrieval technique renders it sensitive to spectral amplitude fluctuations of the employed white-light. Typically, these spectral amplitude modulations are considerably stronger for supercontinua generated in hollow-fibers than the ones obtained from spectral broadening in solid-state materials (see, e.g., Ref. [115]) and are dominated by changes on a wavelength scale larger than the second harmonic bandwidth. Similar to the lock-in detection technique or the radio-frequency heterodyning used for the CEP stabilization of oscillators, the impact of such spectral amplitude modulations on the CEP retrieval can be lessened by encoding the information into a fringe pattern having a periodicity considerably smaller than the spectral amplitude distortions. This is achieved by employing a greater delay between the f and $2f$ components. Hence, also the improvement of noise rejection requires a setup capable of detecting the CEP from the interference of spectral components with non-vanishing delay.

In order to accomplish this goal, the metal prism can be miniaturized and arranged in a row to form an array of reflective Fresnel's biprisms. By placing this array of reflective Fresnel's biprisms in the image plane of the spectrograph as shown in Fig. 2.11 and by matching the periodicity of the spectral interference pattern to the periodicity d of this array, every groove of the grating splits one period of the incident interference pattern into two halves that are reflected to opposing direc-

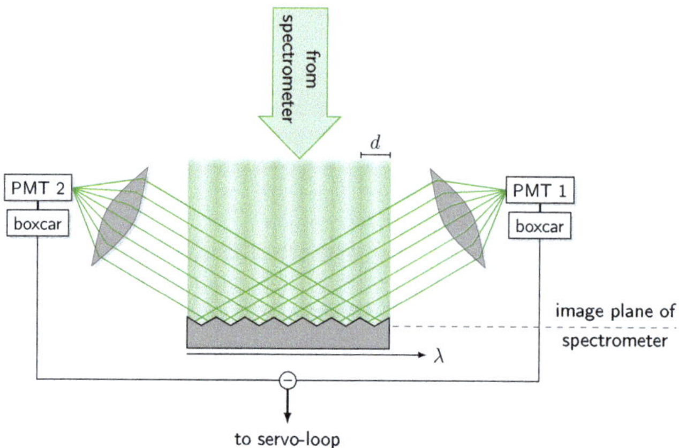

Figure 2.11.: Setup for single-shot carrier-envelope phase retrieval with improved rejection of noise stemming from spectral fluctuations of the supercontinuum. The drawing shows a sagittal cut near the image plane of the spectrometer. The configuration shown here illustrates only schematically the general approach to alternatingly direct half-cycles of the spectral interference pattern onto two different photodetectors and can also be implemented using other splitting schemes.

tions. Hence, the reflected light can be easily focused onto two separate detectors. For the half-meter spectrograph and the experimental setup utilized in the above demonstration of the single-prism-based fast f–$2f$ interferometer, a spatial periodicity of $d \sim 100$ µm and the visibility of approximately 30 spectral interference fringes is achieved by just introducing 1 cm of fused silica into the beam in front of the frequency doubling crystal. Thus, the reduced shot-noise level of the two-detector arrangement can be combined with an improved noise rejection of spectral amplitude modulations by averaging CEP retrieval over several fringes without the need for complicated modifications.

2.3. Concluding Remarks

In this chapter, two important steps to remove the veil of technical noise of CEP stabilization setups have been devised. This allowed analyzing the noise generating mechanisms in more detail.

The systematic comparison of interferometer topologies presented in the first part shows that a compact interferometer design serves to minimize the impact of acoustic noise and air streaks. Compared to the proposed QCP designs, noise rejection may be improved even further by shrinking the unshared path employing, e.g., birefringence [111, 112] for group delay compensation. In addition, the DQCP design can be improved by a monolithic layout in order to render it as noise free as possible. However, the analysis of the topologies shows that, although the interferometers were successively designed to be more robust, the drift like CEP noise component could only be reduced by a factor of 4 compared to the conventional two-path interferometer. Interestingly, also previous Ti:sapphire CEP stabilization studies employing a monolithic f–0 interferometer [54], an octave spanning oscillator [74, 116], or an birefringence based interferometer [112] consistently report such an increase of the PND towards lower frequencies, which typically starts to peak out of the high-frequency plateau in the region between 10 Hz and 100 Hz and increases by one order of magnitude per decade frequency decrease. This indicates that the origin of these long-timescale CEP error contributions might not only stem from the commonly suspected relative interferometer drift. Further investigations on this point, pinpointing one responsible noise source, will be presented in Chapter 4.

In the second part of this chapter, a novel all-analog CEP detection scheme for amplifier laser pulses has been demonstrated. By employing a two-detector arrangement the signal-to-shot-noise ratio is significantly improved compared to the conventional CCD based scheme. The reduction of the shot noise limitation and other technical noise contributions to the extent possible allowed for meaningful single-shot measurement and stabilization of the CEP at 3 kHz repetition rate for the first time. As a result, the analog method clarified the long-standing question on the origin of residual CEP noise in amplifier laser systems. Quite clearly two contributions have been identified in the laser system under study: one glitch-like fast mechanism that currently only appears explainable by fluctuations of the amplifier pump. The remaining noise is found to be inherited from the residual oscillator CEP noise. Hence, further progress in the oscillator CEP stabilization will constitute a major step ahead towards improved sources of high energy CEP-stable pulses for high-field physics and attosecond science.

Chapter 3

Direct Frequency Comb Synthesis with Arbitrary Offset

So far, CEP stabilization of oscillators has been exclusively achieved by a servo-loop approach. In this approach, the f_{CE} beat signal derived from an f–$2f$ interferometer is RF heterodyned with a reference signal and phase coherently locked to it by means of a PLL. Despite its proven utility, however, the servo concept has some distinct drawbacks.

First, generating a pulse train with constant electric field structure, i.e., stabilizing $f_{CE} = 0$, requires further measures since the influence of $1/f$-noise becomes dominant otherwise. Circumventing this limit of the RF heterodyning technique by introducing an AOM into one arm of the f–$2f$ interferometer [58, 117] and artificially shifting the detected signal out of the baseband further complicates the interferometer setup and restricts the interferometer topology to noise susceptible two-path layouts.

Second, since for CEP stabilization the difference between group and phase velocity has to be adjusted intracavity, feedback has to act back on the laser. This feedback, however, may trigger subsequent laser dynamics. Thus, using one of the methods discussed in the introductory chapter frequently causes detrimental dynamic side effects on other laser parameters such as output power, pulse duration or roundtrip time and may cause time-delayed back-action on the CEP drift.

Third, achieving ideal overall CEP stabilization performance requires a careful balance between stability against drop-outs on the one hand and low short-term phase jitter on the other. This becomes particularly intricate as pump lasers for Ti:sapphire lasers are highly complex. Those lasers typically are frequency-doubled Nd-lasers that are pumped by diodes bars in order to achieve an output power of several Watts. This complexity demands cascading of several servo loops controlling different laser parameters for output power stabilization. Furthermore, intertwining analog and digital servo loops are employed for achieving both short- and long-term stability. Under certain external circumstances, however, the intertwining of servo loops may render the combined servo loop unstable in the handover frequency range, which can translate into spikes and glitches in the CEP drift. Robustness against such drop-outs may be achieved using digital phase detectors having a larger unambiguity range than analog ones. This measure, however, comes at the expense of a lower phase resolution. On the contrary, optimization of the short-term phase jitter is subject to a basic restriction: stable operation of the PLL requires a phase margin greater than zero, which limits achievable servo-loop bandwidths even in carefully optimized setups [72].

In this chapter a conceptually different self-referenced feed-forward approach to CEP stabilization is demonstrated. The proposed method neither requires any servo-loop nor compromises laser performance since the central element is placed outside the laser and the oscillator is kept free-running. The method enables direct synthesis of frequency combs with freely selectable comb offsets. The residual CEP jitter is shown to surpass the precision of the servo-loop method by a factor of five. Furthermore, the performance is found to be limited only by detection shot noise with further improvement at hand.

The results presented in this chapter have been published in Ref. [118]. They constitute a considerable performance improvement of the earlier measurements in the thesis of Christian Grebing [103]. These improvements are obtained by a more robust layout of the f–$2f$ interferometers using a QCP arrangement. In addition, the sensitivity of this novel CEP stabilization scheme to intensity noise of the pump laser is studied for the first time by comparing the results obtained with two different laser models. In addition, a previously unrecognized source of phase error in this scheme is identified and a method for its circumvention is proposed.

3.1. Concept for Direct Frequency Comb Synthesis

The self-referenced feed-forward scheme bases on the modification of properties of the beam with the help of an acousto-optical frequency shifter (AOFS), as illustrated in Fig. 3.1. A transducer, which is attached to one facet of a bulk medium, induces mechanical strain inside the medium of the AOFS. The mechanical strain forms an acoustic wave, which propagates through the medium and harmonically alters the refractive index via the photoelastic effect [119]

$$\Delta n(\boldsymbol{r}, t) = \delta n \cos\left[2\pi f_{\mathrm{ac}}t - \boldsymbol{K}_{\mathrm{ac}}(t)\boldsymbol{r} + \varphi_{\mathrm{ac}}\right] . \tag{3.1}$$

Here δn is the amplitude of the refractive index modulation and $\boldsymbol{K}_{\mathrm{ac}}$ is the wave vector of the acoustic wave. The spatial refractive index modulation acts as a transmission grating diffracting incident light into higher diffraction orders.

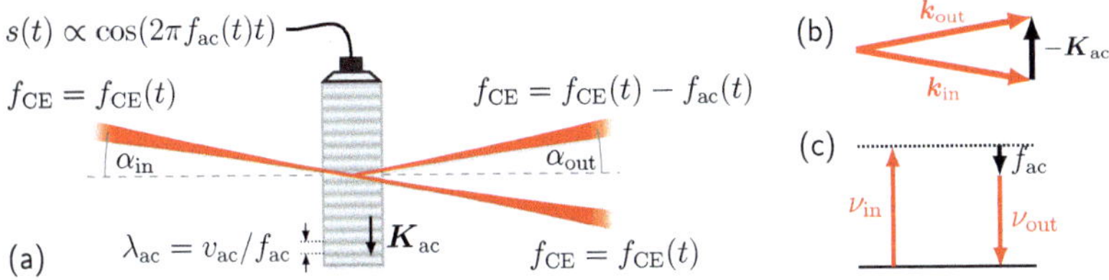

Figure 3.1.: (a) Illustration of the geometric arrangement for Bragg diffraction from traveling acoustic waves inside the frequency shifter leading to a frequency downshift by f_{ac} ($m = -1$). Bragg reflection occurs when $\alpha_{in} = \alpha_{out} = \theta_{B}$. (b) and (c) Corresponding wavevector and energy diagram, respectively.

Since diffraction into several higher diffraction orders is unwanted in the application that follows, the interaction length between acoustic-wave and optical beam is made large enough by increasing the thickness d of the sound column. Then contributions of the higher orders cancel each other out, and Bragg diffraction of the incident light into one order occurs [120]. This Bragg regime is characterized in terms of the Klein-Cook parameter $Q = dK_{ac}^2/k$ and the Raman-Nath parameter $v = 2\pi d\delta n/(\lambda c)$ (k wavevector and λ wavelength of the incident light, c vacuum speed of light) by

$$Q \gg 1, \qquad Qv \gg 1. \tag{3.2}$$

As known from solid-state physics, for a given wavelength, Bragg diffraction is only encountered at one specific angle θ_{B} determined by

$$\theta_{B}(\lambda) = \arcsin\left(\frac{m\lambda}{2\lambda_{ac}}\right). \tag{3.3}$$

Here, $m = \pm 1$, depending on whether the component of the incident light $\boldsymbol{k}_{in}$ is antiparallel or parallel to $\boldsymbol{K}_{ac}$, respectively. As depicted in Fig. 3.1(b), the wavevector of the reflected beam is determined by $\boldsymbol{k}_{out} = \boldsymbol{k}_{in} + m\boldsymbol{K}_{ac}$, i.e., depending on the sign of m, one phonon wavevector $\boldsymbol{K}_{ac}$ is added to or subtracted from the incident wavevector. Equivalently, the transducer driver frequency f_{ac} enters with a positive or negative sign in the energy diagram shown in Fig. 3.1(c).

Consequently, for the downshift case illustrated in Fig. 3.1(a), the frequency comb of the Bragg reflected pulse train reads

$$\nu_n = f_{CE}(t) - f_{ac}(t) + n\,f_{rep}. \tag{3.4}$$

Employing RF synthesis of the driver signal

$$f_{\mathrm{ac}}(t) = f_{\mathrm{synth}} + f_{\mathrm{CE}}(t) \tag{3.5}$$

enables the generation of frequency combs

$$\nu_n = f_{\mathrm{synth}} + n\,f_{\mathrm{rep}} \tag{3.6}$$

with freely selectable offset frequency f_{synth}. Hence, the carrier-envelope frequency (CEF) of the pulse train leaving in the first diffraction order is stabilized to f_{synth}.

However, not only the frequency is stabilized. By applying coupled mode theory to the Bragg reflection problem [121] it is readily shown, that variations of the spatial phasing φ_{ac} of the refractive index modulation (3.1) translate into phase changes of the diffracted beam. Since the phase of the f–$2f$ beat signal resembles the CEP up to an unknown constant, CEP stabilization is achieved when a driver signal according to Eq. (3.5) is used.

In the special case of choosing $f_{\mathrm{synth}} = 0$, the frequency comb is stabilized to the important case of zero offset, such that the stabilized pulse train in the first diffraction order consist of pulses with identical electric field structure.

3.2. Experimental Verification

Fig. 3.2(a) shows the setup that is used to demonstrate the self-referenced feed-forward scheme introduced in the last section. As a laser source, a sub-10-fs oscillator is employed, which is pumped by a diode-pumped solid-state laser (Spectra Physics Millennia). The output of the oscillator is spectrally broadened in an MSF to obtain octave coverage. The spectrally broadened beam is recollimated and focused into the AOFS (Brimrose QZF-70-10-.800), which is 2.7 cm thick and made of quartz. For convenience, the CEF is measured with an f–$2f$ interferometer [right-hand-side box in Fig. 3.2(a)] in the zero-order output, which alleviates the need for an additional beam splitting in front of the setup. This IL interferometer serves to synthesize the AOFS drive signal. The light diffracted into the first order is directed into a second OOL f–$2f$ interferometer [left-hand-side box in Fig. 3.2(a)] for an independent characterization of the residual CEP jitter in the CEP stabilized beam. Both interferometers are built in a quasi-common path geometry [102] to minimize spurious noise contributions by, e.g., a relative drift of interferometer arm lengths. The OOL interferometer is designed to compensate for the angular dispersion in the first diffraction order. As shown in Fig. 3.2(b) and (c), the signal-to-noise ratios obtained in the IL and the OOL interferometer are 40 and 30 dB (RBW 100 kHz, VBW 300 kHz), respectively.

As the measurement in the zeroth order remains unaffected by the resulting frequency shift in the first order, this method has a feed-forward character. By using

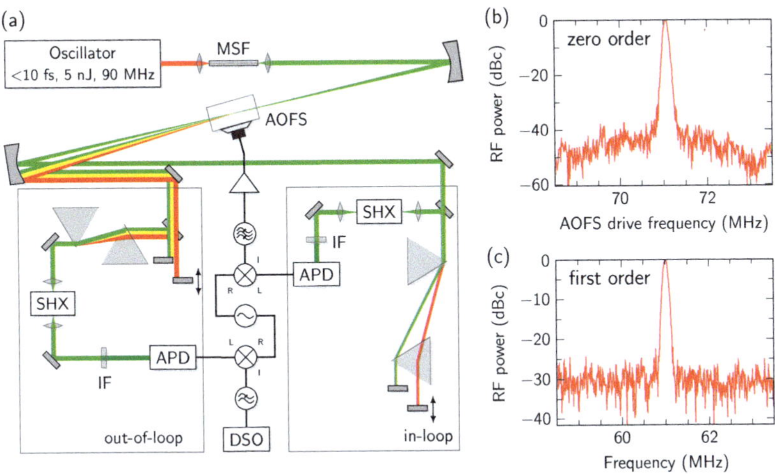

Figure 3.2.: (a) Experimental setup used for out-of-loop CEP stabilization characterization of the feed-forward scheme. MSF: micro-structured fiber, PPLN: periodically poled lithium niobate, IF: interference filter, APD: avalanche photodiode. (b) and (c) f_{CE} beat notes that are obtained in the zero and the first diffraction order beam, respectively (100 kHz RBW, 300 kHz VBW).

an quartz AOFS with 70% diffraction efficiency, more than half of the laser power is available for applications of the CEP stabilized pulses. Therefore, the power ratio between the beam that is used to generate the CEF signal and the beam that is available for application in this scheme is identical to the typical power ratios employed in the servo-loop approach. The AOFS device is optimized for operation at 70 ± 10 MHz, and the maximum 70% diffraction efficiency are reached at $+38$ dBm drive power. From manufacturer data for the velocity of sound $v_{ac} = 5960$ m/s and for the diffraction efficiency, the Klein-Cook parameter is computed to have the value of $Q \approx 17$ and the Raman-Nath parameter to have the value of $v \approx 2$ (details can be found in Section 3.3), such that conditions (3.2) for Bragg reflection are fulfilled. This is backed by the observation of a single diffraction order in the experiment.

The electronic phase lag caused by the travel time of the acoustic wave from the transducer to the interaction zone has been minimized to the extent possible. Mechanical construction of the AOFS allows for a minimum distance between laser beam and acoustic actuator of approximately 2 mm, which translates into an estimated feed-forward loop bandwidth of ≈ 1.5 MHz. This is more than one order of magnitude larger than the best loop bandwidths of ≈ 100 kHz reported for optimized feedback servo loops [72].

The RF heterodyne CEP jitter detection scheme is not applicable to zero-offset frequency combs, since at $f_{CE} = 0$ positive CEF deviations cannot be distinguished from negative ones. This ambiguity can only by avoided by switching to a homodyne detection setup and by placing an additional detector in the second output port of an MZ interferometer. Analyzing the relative amplitude fluctuations on the two detectors then allows for CEP retrieval [122]. However, increased $1/f$-noise contributions in the electronics, spurious noise of the MZ topology, and non-ideal balancing of the two detectors renders this setup less sensitive than RF heterodying. In order to avoid these drawbacks, the signals are shifted out of baseband by employing RF synthesis with a fixed synthesizer frequency $f_{\mathrm{synth}} = 60$ MHz derived from a quartz-stabilized RF oscillator. Heterodyning this reference with the independently measured OOL beat signal at f_{synth} [see Eq. (3.6)] gives access to residual CEP jitter of the stabilization.

Figure 3.3(a) shows a 5 s long time series of the residual OOL CEP jitter achieved with this setup. The jitter is recorded by a digital sampling oscilloscope at a sample rate of 5 MSa s^{-1}. Additionally, a second measurement was recorded with the f and $2f$ components in the OOL interferometer being temporally mismatched to suppress interference. As this signal does not carry any phase information, but the same number of photons is detected, the resulting heterodyne signal allows estimation of the shot-noise induced CEP detection limit. The Fourier transforms of both signals are shown in Fig. 3.3(b). These curves indicate that the residual phase noise is essentially limited by detection noise in the range from 30 Hz to the sampling limit of 2.5 MHz, with only relatively few discrete frequencies appearing significantly above detection noise floor. Over 5 decades, therefore, the performance of the self-referenced feed-forward CEP stabilization scheme is determined by detection shot noise. This situation is in strong contrast to previously reported CEP stabilization performance [36, 54, 69, 73], where a broad acoustic band was consistently reported that exceeded the detection limit by some 25 dB, peaking in the 10 to 1000 Hz region. Similar noise contributions have also been observed in the interferometer topology comparison presented in Section 2.1. Moreover, it is also striking that the spectral analysis does not report any prevalent noise contributions in the 100 kHz range, which have previously been observed in servo stabilized lasers and attributed to the multi-mode concept of the pump laser employed [123]. Given their relatively high frequency above the loop bandwidth, such phase noise contributions are impossible to equalize in traditional servo-loop concepts. This limitation does not play a role in the feed-forward concept, which is only limited by the bandwidth of the AOFS of some MHz. At frequencies below ≈ 10 Hz, slow dephasing between the OOL and IL signal is observed. To some extent, this dephasing stems from pump power fluctuations, which will be discussed in the next chapter. In addition, a relative

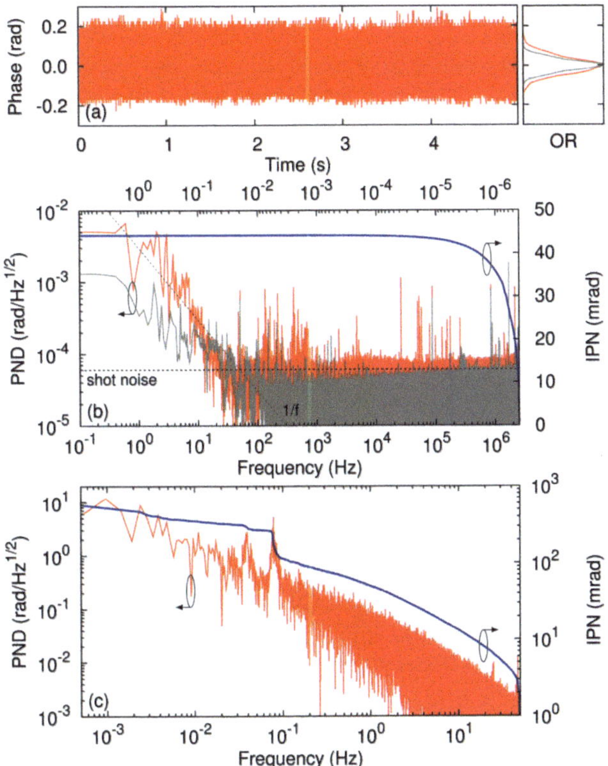

Figure 3.3.: Performance of the self-referenced feed-forward carrier-envelope phase
stabilization scheme. (a) 5 s long time series of the out-of-loop phase
error (red) and the distribution of occurring phase values (OR) for
the out-of-loop phase error and the shot-noise induced phase error
(gray). (b) Phase noise density (PND) of the out-of-loop phase error
(red), shot-noise error (gray), and the out-of-loop integrated phase
noise (IPN, blue). (c) Out-of-loop PND (red) and IPN (blue) of a
35 min long measurement.

drift of the two interferometers may also contribute. Note that no means were undertaken to stabilize the CEF of the free-running laser oscillator in any way. The slow dephasing is relatively minor, amounting to an RMS excursion of 570 mrad in a 35 min observation time [Fig. 3.3(c)]. This drift-like $1/f$ noise is estimated to exceed detection noise at frequencies below ≈ 100 Hz, which also compares favorably with previously reported stabilization behavior [54, 69].

The total integrated phase noise (0.2 Hz to 2.5 MHz) of the data in Fig. 3.3(b) amounts to 45 mrad, which corresponds to a temporal jitter of about 20 as. At higher frequencies, the performance of the feed-forward scheme appears to be only limited by detection shot-noise, which by itself already amounts to 30 mrad. It is striking that residual phase noise exceeds detection noise by a factor of only about $\sqrt{2}$ in the high frequency region. By assuming Gaussian statistics and fitting Gaussians to the distribution of occurring phase values shown in the right graph of Fig. 3.3(a), it is estimated that the true underlying residual phase noise jitter amounts to ≈ 30 mrad, corresponding to a timing jitter of only 12 as. While this value corrects for noise in the OOL interferometer, it does not account for shot noise in the IL interferometer. The bottleneck of the current detection scheme is the frequency-doubling in the $2f$ arm. Generation of a supercontinuum with stronger contributions in the near infrared or optical amplification prior to doubling could serve to further improve the signal-to-noise ratio of the setup. This measure promises to further push residual jitter to even lower values.

For comparison, Table 3.1 lists the OOL CEP stabilization results that have previously been achieved with the servo-loop approach. As the measurements employing the monolithic f–0-interferometer in Refs. [54, 73] and the octave spanning oscillator [74] cover the complete frequency range of the feed-forward measurement, phase errors can be recalibrated to the frequency range resolved in the feed-forward measurement enabling an unbiased comparison of the results. By digitization of the integrated phase noise curves reported in these articles, the residual CEP jitter in between 0.2 Hz and 2.5 MHz is calculated to be 150 mrad [54], 75 mrad [73], and 70 mrad [74]. The origin for the improved residual CEP jitter in the second paper of Fuji et al. [73] is owing to a > 20 dB enhancement of the beat note signal-to-noise ratio [54]. As explained in the Appendix A.2, such an signal-to-noise improvement results in more than one order of magnitude decrease in the impact of the detection noise floor. This fact points out that only the comparison of measurements made with similar signal-to-noise ratios are meaningful. In consequence, a $4-5$ times better CEP stability is obtained using the feed-forward scheme compared to the results of the feedback scheme employing beat notes with equal signal-to-noise ratios.

Of special interest is also the comparison with the performance of the optical self-referencing technique demonstrated for high energy pulses [52]. In this approach difference-frequency generation between two combs with identical offsets $f_{CE} \neq 0$ in an optical parametric amplifier is used to produce zero-offset combs. The energy diagram of the parametric process resembles the one of the feed-forward approach in

Reference	RMS jitter	Frequency range	OOL SNR
Fuji et al. [54]	295 mrad[a]	0.48 mHz $-$ 35 MHz	> 35 dB
Fuji et al. [73]	100 mrad[b]	0.2 mHz $-$ 35 MHz	> 55 dB
Crespo et al. [74, 116]	75 mrad[c]	2 mHz $-$ 10 MHz	55 dB
Witte et al. [123]	152 mrad	250 Hz$- < 13$ MHz	$-$
Fortier et al. [69]	720 mrad	976.5 μHz $-$ 102.4 kHz	$-$
this work	$30 - 45$ mrad	0.2 Hz $-$ 2.5 MHz	$\simeq 30$ dB

[a] $\simeq 150$ mrad over 0.2 Hz–2.5 MHz
[b] $\simeq 75$ mrad over 0.2 Hz–2.5 MHz
[c] $\simeq 70$ mrad over 0.2 Hz–2.5 MHz

Table 3.1.: Comparison of the CEP stabilization results reported in literature with the results obtained for the feed-forward approach. In order to render the comparison independent of the resolved frequency ranges, the phase errors integrated over the frequency range of the feed-forward measurement are quoted in the footnotes. Note that not only frequency regions of measurements are different, but that setups also vary: while Refs. [54, 73] used self-phase modulation in the difference-frequency generation crystal for spectral broadening up to an octave, Refs. [69, 123] employed spectral broadening in two independent micro-structured fibers, such that additional noise from the fiber might couple into the results. Furthermore, in Refs. [74, 116] an already octave spanning oscillator was used and only an in-loop characterization was performed. According to the argumentation of the authors, this truthfully characterizes the residual CEP jitter since there is no nonlinear interaction process corrupting an out-of-loop characterization for an octave spanning oscillator. Nevertheless, the comparison shows that a factor of $4 - 5$ improvement is yielded compared to Ref. [54], which analyzed the CEP jitter with a comparable signal-to-noise ratio of the beat note.

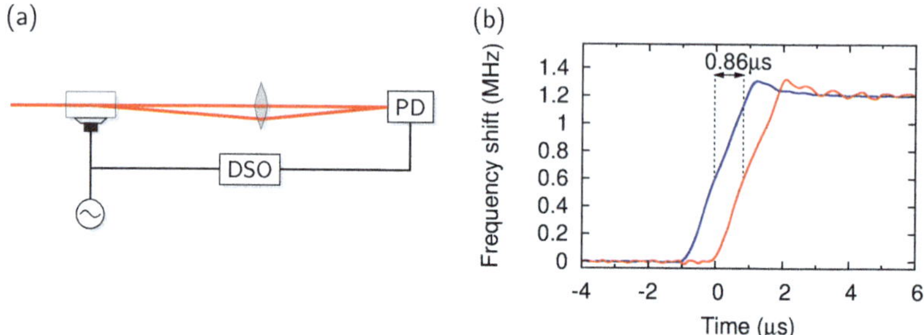

Figure 3.4.: (a) Setup for the measurement of the travel-time induced delay of the frequency modulation (PD: photodiode; DSO: digital storage oscilloscope). (b) Delay between the frequency modulation of the AOFS drive signal (blue) and the frequency modulation measured in the beat of the first diffraction order light with the zero order light (red). The ripples on the red trace originate from the lower signal-to-noise ratio (≈ 25 dB) of the beat detection in combination with the subsequent numerical frequency demodulation.

Fig. 3.1, with the acoustic wave being replaced by an optical signal beam. For such an approach, Baltuška et al. achieved an RMS CEP stability of 180 mrad averaged over 100 laser shots and attributed the residual CEP jitter to intensity-to-phase coupling that probably results from the intermediate supercontinuum generation process or depletion of the optical parametric amplifier's pump beam [124, 125]. Using an AOFS as self-referencing element such amplitude-to-phase coupling issues are avoided, and a clearly superior CEP stability is achieved.

Bandwidth of the Feed-Forward Scheme

As was argued above, the bandwidth of feed-forward concept is determined by the travel time of the acoustic wave from the actuator to the interaction zone, which should permit the achievement of bandwidths > 1 MHz. In order to check the above derived estimate, the bandwidth of the feed-forward loop is measured using the setup shown in Fig. 3.4(a). For this measurement, the AOFS is driven with the frequency modulated output of an RF generator. The beat signal between the light leaving the AOFS in the first and in the zero order is recorded with an oscilloscope, and the frequency modulation is demodulated via phase retrieval with the Takeda algorithm [126]. The results of the measurement presented in Fig. 3.4(b) reveal that the travel-time induced delay is about 0.9 µs, i.e., very close to the above estimate, enabling feed-forward bandwidths of > 1 MHz. Therefore, the feed-forward concept

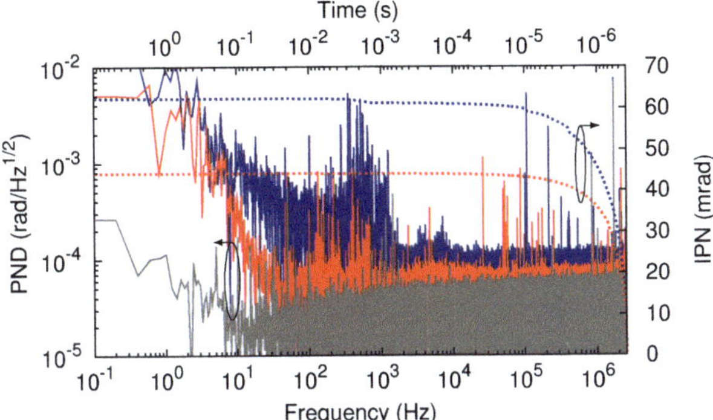

Figure 3.5.: Comparison of the feed-forward CEP stabilization performance obtained with different pump lasers; red: multi-mode pump laser [Spectra Physics Millennia; the same data as in Fig. 3.3(b)]; blue: single-mode pump laser (Coherent Verdi); gray: detection noise level for the Coherent Verdi measurement.

exceeds by far the best published bandwidths in a feedback loop [72], promising elimination of noise sources far beyond the capabilities of conventional schemes.

Impact of Pump Laser Intensity Noise

Previous investigations showed a pronounced dependence of the servo lock CEP stabilization performance on the type of laser used for pumping of the Ti:sapphire oscillator [72, 123]. Since the amount of RIN is directly written onto the CEP within the oscillator's cavity bandwidth, this dependence results from the different frequency distribution of the pump laser's RIN. As a result, CEP stabilization with the servo lock approach has mainly been achieved with the Verdi from Coherent, which is a single-longitudinal-mode laser. To a lesser extent, CEP stabilization has been attained using the Millennia from Spectra Physics. Typically, however, the former gave significantly better results which has been attributed to the multi-mode concept of the Millennia [123]. Furthermore, comparative RIN measurements showed a higher RIN for the multi-mode laser at essentially all frequencies above 20 Hz [72].

Figure 3.5 shows a comparison of the feed-forward CEP stabilization of the same Ti:sapphire oscillator employing these two pump lasers. The measurements have been carried out on two consecutive days, and care has been taken to obtain identical mode-locking behavior, as indicated by nearly identical spectra. Quite surprisingly, the single-mode laser displays marginally worse CEP stabilization performance with

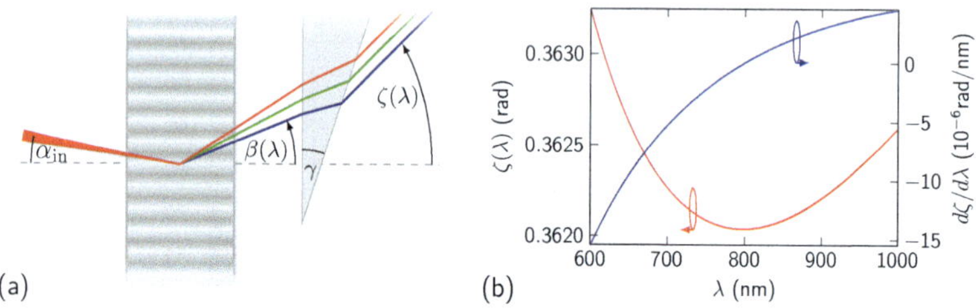

Figure 3.6.: Compensation of the angular dispersion of the acusto optical frequency shifter by insertion of a wedge. (a) Geometrical arrangement of the wedge. (b) Calculated residual angular dispersion for a fused silica wedge having an apex angle of $\gamma = 34°$.

an RMS phase jitter of 68 mrad. The IPN reveals that most of the difference is already accumulated for frequencies > 1 kHz, which stems from slightly different signal levels achieved in the f–$2f$ interferometer. Apart from the impact of the increased detection noise level, small steps in the IPN are observable at 400 kHz, 100 kHz and $300 - 700$ Hz. A comparison with the PND curves shows that these steps are accompanied by stronger contributions observable in the single-mode pump laser measurement. This discrepancy is most obvious in the frequency region from 10 Hz to 1 kHz.

While the exact differences seen in such comparative measurements might also depend on the manufacturing tolerances within one model series or on the degradation of the laser diodes, this study shows, nevertheless, that CEP stabilization using the feed-forward concept is obtained with both laser concepts. Furthermore, both the single-mode and the multi-mode laser allowed for achieving extremely good residual CEP jitters below the best reported values obtained with the traditional servo lock approach.

3.3. Further Improvements

Angular Dispersion of the Diffracted Beam

The acoustic waves inside the AOFS act as a transmission grating for the incident optical beam. Thus, different wavelengths experience different diffraction angles, such that the light in the first diffraction order is angularly dispersed: $\alpha_{\text{out}} = \alpha_{\text{out}}(\lambda)$. This angular dispersion is ideally compensated for by placing an element offering the same but negative angular dispersion directly behind the AOFS. A very convenient compensation realization is to use a double pass through the AOFS by retroreflecting the first diffraction order beam [127]. While this allows for ideal compensation of the angular dispersion, it requires frequency division of the f_{CE} beat signal by a factor of two for CEP stabilization and it also increases the losses. Another compensation method is to employ a sequence consisting of a collimating mirror and a prism pair, as has been used in the experimental demonstration of the feed-forward approach, or inserting a wedge directly behind the AOFS. The latter method is depicted in Fig. 3.6(a). Calculation of the wavelength dependence of the complete diffraction angle

$$\zeta(\lambda) = -\gamma + \arcsin\left\{ n(\lambda)\sin\left[\gamma + \arcsin\left(\frac{\sin\beta(\lambda)}{n(\lambda)}\right)\right]\right\} \tag{3.7}$$

with

$$\beta(\lambda) = \arcsin\frac{2\lambda - \lambda_c}{2\lambda_{ac}} \tag{3.8}$$

allows for optimizing the apex angle of the wedge. Figure 3.6(b) shows the result of such a calculation and corroborates that a fused silica wedge with 34° apex angle compensates for the angular dispersion to a satisfactory degree.

Apart from the angular dispersion, the absolute diffraction angles changes, too, when the AOFS driver frequency drifts. For the AOFS used the change of the diffraction amounts to $\Delta\alpha_{out} = 1.3$ µrad/MHz, such that the electronical AOFS bandwidth of ±4 MHz results in an maximum angle deviation of ±5%. If the passive f_{CE}-stability of the laser is not good enough to cope with the requirements for the application of the CEP stabilized beam, the angular beam pointing can be circumvented by a simple and not necessarily fast frequency lock of f_{CE}. Such a lock can, e.g., be accomplished using intracavity wedges, which are used for coarse CEF adjustment in nearly every CEP stabilized laser. Furthermore, the double pass configuration [127] mentioned earlier compensates for this beam pointing, too.

Impact on the Pulse Duration

Placing the CEP stabilizing element behind the oscillator, is one of the main advantages of the feed-forward approach. By doing so the oscillator is not affected by the action of the CEP stabilization, and triggering of artificial laser dynamics is prevented. As a result, however, the ultrashort laser pulse has to transverse the additional material of the AOFS, which causes a temporal stretching of the pulses. As in CPA laser systems pulses are stretched prior to amplification anyway, this is not an issue if the AOFS is used for CEP stabilization of such systems. The temporal stretching of the pulse is also no hindrance for frequency metrology provided all required spectral components are already contained inside the incident spectrum. However, subsequent nonlinear processes requiring zero group-delay such as spectral broadening or frequency conversion suffer from the stretching.

For the AOFS used in the experiment, a group-delay-dispersion of ≈ 1000 fs^2 and a stretching of the pulse duration[1] to ≈ 300 fs is introduced by the 2.7 cm-long quartz medium. The method of choice for compensation of such an amount of GDD is to use several bounces off chirped mirrors. While this is a very convenient method for 10 fs pulses, losses may not be tolerable anymore if even shorter pulses are used and if the chirped mirrors are driven to the extreme in terms of bandwidth. Then, a prism compressor or a combination of prisms and chirped mirrors may turn out as a better alternative. It is important to note here, however, that, in contrast to the servo-lock approach, pulse broadening is not subject to dynamic changes.

In addition, the spectral bandwidth is reduced for the light that is diffracted into the first diffraction order. This is due to the fact that the wavevector diagram shown in Fig. 3.1(b) holds only for the Bragg wavelength. For all other wavelengths, however, a wavevector mismatch $\Delta \boldsymbol{k}$ appears and renders the process non-phase-matched. The spectral dependence of the diffraction efficiency is described by [120]

$$\eta(\lambda) = \left(\frac{v}{2}\right)^2 \mathrm{sinc}^2 \sqrt{\left(\frac{v}{2}\right)^2 + \left(\frac{K_{\mathrm{ac}}\Delta\alpha d}{2}\right)^2} \, , \tag{3.9}$$

where $\Delta\alpha(\lambda) = \alpha_{\mathrm{in}} - \theta_B(\lambda)$ is the angle offset from Bragg incidence at a given wavelength. Using (3.9), neglecting the wavelength dependence of the elasto-optic tensor, and using the manufacturer given data for the diffraction efficiency η_r at reference wavelength λ_r, the Raman-Nath parameter is calculated via[2]

$$v(\lambda) = \frac{2\lambda_r}{\lambda} \arcsin \sqrt{\eta_r} \, . \tag{3.10}$$

Insertion of the values of the experimental setup yields the diffraction efficiency curve presented in Fig. 3.7(a). The bandwidth reduction that is caused by the spectral dependence of the diffraction efficiency leads to an increase in the Fourier-limited pulse duration [see Fig. 3.7(b)], which becomes increasingly severe when the incident pulse duration is significantly below ≈ 7 fs. In order to make this phase matching induced pulse broadening less pronounced the relative magnitude of the difference angle term in (3.9) has to be made smaller compared to the Raman-Nath term. This can be achieved by either enhancing the value of the Raman-Nath parameter, i.e.,

---

[1]This estimation assumes an incident Fourier-limited Gaussian beam with $\tau_0 = 10$ fs full-width half maximum pulse duration. Such a pulse is broadened to $\tau_{\mathrm{out}} = \tau_0\sqrt{1 + 4\ln 2 \cdot |\mathrm{GDD}|/\tau_0^2}$ [82].

[2]For the quartz shifter used in the experiment, the manufacturer specifies $\eta_r = 70\%$ at $\lambda_r = 792$ nm and a sound column thickness of 22 mm.

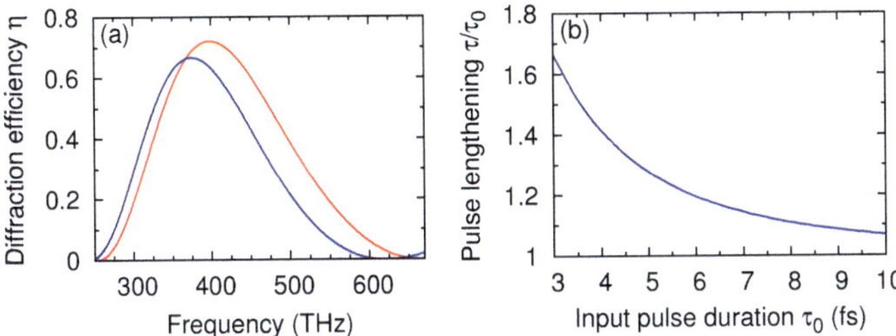

Figure 3.7.: Illustration of the influence of phase matching. (a) Diffraction efficiency vs. frequency of light for Bragg angle incidence at specification wavelength (red), $\alpha_{\text{in}} = \theta_B(792\,\text{nm})$, and for symmetry optimized incidence angle of $\alpha_{\text{in}} = \theta_B(840\,\text{nm})$ (blue). (b) Pulse lengthening introduced by bandwidth reduction for optimized incidence angle. For this calculation a Fourier-limited Gaussian temporal shape has been assumed, and pulse durations are measured by the intensity full-width half maximum.

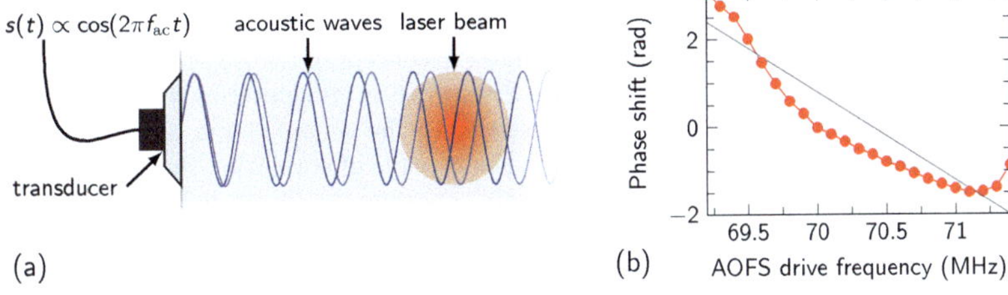

Figure 3.8.: (a) Illustration of the different phasings of the acoustic waves of different frequency with respect to the center of the laser beam, which results from the finite distance of the interaction region from the transducer. Similarly, every phase lag in the driving electronics circuit influences the interaction with the laser beam. Hence, introducing a suitable phase shift in the electronics allows for compensation of the drive frequency dependent phase response of the acousto-optic frequency shifter by bringing the fixed point of the acoustic waves to coincidence with the center of the incident beam. (b) Measurement of the phase shift (red) in the acousto-optic frequency shifter position that was used to obtain the reported carrier-envelope phase stabilization. Approximate linear slope expected for the travel distance introduced phase response calculated via Eq. (3.11) (gray).

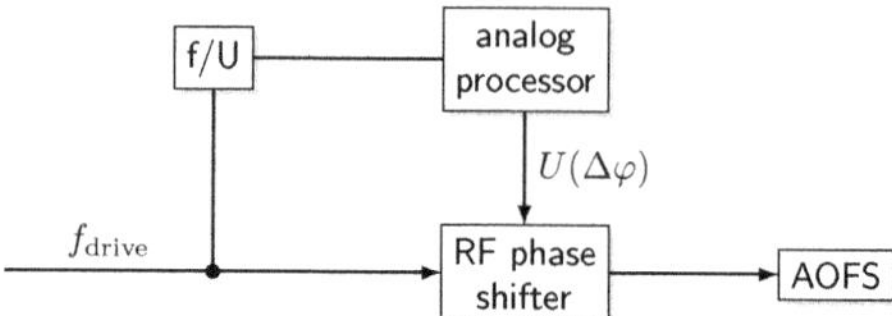

Figure 3.9.: Setup for compensation of the phase response of the electronics. To this end, the frequency of the driver signal is converted to a voltage by an f/U-converter. The resulting voltage is processed with an analog processor that, after a calibration, generates an appropriate control signal $U(\Delta\varphi)$ for the RF phase shifter. Finally, the input signal is shifted by $\Delta\varphi$ using the RF phase shifter and is used as new driver signal for the AOFS.

increasing the RF power in order to utilize a higher refractive index modulation, or lowering the driver frequency f_{ac}, which is the more effective method.

Influence of the Phase Response of the Electronics

As has been discussed in the introduction to the concept of the feed-forward scheme, the phase of the light diffracted to the first order is influenced by the phasing of the refractive index modulation with respect to the center of the incident beam. While precisely this fact renders this scheme capable of stabilizing the CEP, it also renders it sensitive to changes of the relative phasing. As illustrated in Fig. 3.8(a), these changes may result from changes in the driver frequency f_{ac} due to the travel distance induced phase response. At a distance Δx from the transducer, the phasing φ_{ac} of the acoustic wave is given by

$$\varphi_{\mathrm{ac}} = \frac{2\pi f_{\mathrm{ac}}\Delta x}{v_{\mathrm{ac}}}. \tag{3.11}$$

For $\Delta x \approx 2$ mm this results in a slope $d\varphi_{\mathrm{ac}}/df_{\mathrm{ac}} = 2$ rad/MHz. As can be seen from experimental data plotted in Fig. 3.8(b), this estimation approximates the measured progression quite well. The deviation from a linear slope originates from contributions of the electronics, which probably stem from the LC-circuit that is applied for resonant enhancement of the transducers drive signal.

In future applications the influence of the changing phasing can be circumvented by implementing a RF phase shifter into the electronics as shown in Fig. 3.9. After a calibration, the phase shift introduced by the RF phase shifter can be tuned according to the offset from a frequency setpoint, such that the fixed point of the refractive index modulation comes to lie at the center of the beam for every driver frequency.

Furthermore, implementation of the RF phase shifter also adds functionality: in applications of the feed-forward CEP stabilization in CPA laser systems it enables modulating the CEP behind the amplifier section by adding an appropriate phase shift to the AOFS drive signal using the RF phase shifter. As this does not interfere with the compensation of the oscillator's CEP drift and is only limited by the travel time induced bandwidth, feeding the slow-loop's error signal back to the RF phase shifter enables CEP stabilization at the full repetition rate of the amplifier.

3.4. Concluding Remarks and Outlook

In summary, a novel method for the generation of self-referenced optical frequency combs has been devised in this chapter. In contrast to previous servo-loop approaches, the novel CEP stabilization method is of feed-forward type. As a result, the bandwidth constraint dictated by the maintenance of a feedback is removed. Furthermore, no intervention into the laser is made, such that no transient laser dynamics are triggered by the CEP stabilization. Compared to servo-loop stabilizations, the unprecedented residual jitter of only 45 mrad mainly results from a reduction of noise contributions in the acoustic frequency range between 10 and 1000 Hz. The characterization of the residual CEP jitter was shown to be limited by the shot noise of the beat detection. Thus, amplification of the low light levels in the f–$2f$ interferometer or utilization of already octave spanning oscillators promise to push the residual CEP jitter to even lower values.

Outlook

The stabilization performance achieved in this demonstration will be interesting to the field of frequency metrology. Although in frequency metrology it theoretically suffices to know the f_{CE}-beat signal at every instance of time [41], those measurements typically employ CEP stabilized lasers due to the gain in precision resulting from a more stable comb. Using the traditional feedback servo-loop concept, only low-frequency noise < 100 kHz can be compensated such that the short-term instability of the laser mandates long averaging times in order to obtain low uncertainties [128]. This issue has led to the development of the transfer concept [128], in which the measurement is designed to minimize the impact of fluctuations of the frequency comb on the accuracy of the measurement. In addition, high-frequency technical noise is suppressed in this concept by tracking the beat notes with fast all-electronic PLLs having a bandwidth of ≈ 1 MHz [128]. The feed-forward CEP stabilization scheme offers a CEP stabilization bandwidth that is equal to the bandwidth of the transfer concept and that, in principle, could be enhanced by using electro-optic frequency shifting. Although the implementation of an electro-optic frequency shifter for pulse trains with a MHz repetition rate seems challenging, the combination of the transfer concept with a bandwidth extended feed-forward stabilization may therefore enable a further progress in noise rejection such that, e.g., the investigation of the stability of fundamental constants [17] will come closer into reach.

In this respect it is interesting to note that, as fiber versions of AOFS are commercially available, the feed-forward scheme is easily implemented in ultrafast fiber lasers that are traditionally used in frequency metrology due to their improved long term stability. Furthermore, using fiber based f–$2f$ interferometers [107] contributions of additional interferometer noise can be circumvented, too. This should yield a highly reliable CEP stabilized laser as no lock has to be maintained in the feed-forward technique such that the stabilization will immediately reestablish after a potential brief interruption of the input signal.

The feed-forward approach seems to be ideal for the CEP stabilization of, e.g., chirped-pulse oscillators [129]. In this oscillator concept, a chirped pulse passes through the laser crystal, which enables generation of pulses with higher pulse energies without encountering nonlinearity induced mode-locking instabilities limiting the energy scalability in conventional oscillators. In turn, however, the decreased nonlinear interaction probably permits the utilization of pump power modulation for CEP stabilization since the accessible CEP modulation range is expected to be too small. In contrast, the feed-forward approach does not suffer from this constraint as it does not rely on the exploitation of intracavity nonlinear propagation effects.

Moreover, attosecond physics will benefit from the improved CEP stability in two ways. First, to the end of generating shorter pulses, ever steeper electric field gradients are employed in order to achieve broader cut-offs in high-harmonic spectra. As a result, the attosecond pulse generation process is rendered even more susceptible to fluctuations in the driving electric field translating into a jitter of the attosecond pulse duration. These fluctuations are minimized with the feed-forward CEP stabilization. Second, as attosecond pulses are emitted at the zero crossings of the driving electric field, fluctuations of the CEP induce variations of the emission timepoints. Hence, the CEP jitter causes a jitter of the pump-probe delay in femtosecond-pump attosecond-probe experiments, which is on the order of the shortest attosecond pulse duration of 80 as [44] in feedback stabilizations. This compromises the obtainable temporal resolution. For the first time, the feed-forward concept enables reducing the pump-probe delay jitter to values significantly lower than the attosecond pulse durations. In consequence, the superior timing control offered by the feed-forward approach will facilitate the accurate generation of even shorter attosecond pulses and may therefore contribute to the investigation of a large variety of transient processes on the attosecond time scale.

Chapter 4

Sub-Shot-Noise Carrier-Envelope Frequency Jitter in Kerr-Lens Mode-Locked Lasers

The feed-forward CEP stabilization results presented in the previous chapter raise the question what ultimately limits the quality of any CEP stabilization scheme.

Apart from technical limitations such as the beat note's finite amplitude, the per-roundtrip CEP shift is subject to fundamental limitations. Up to now, the noise added by spontaneous emission to the electric field traveling inside the cavity has been identified as the fundamental limit for the achievable CEP jitter [130, 131]. However, in the measurements presented in this thesis, such an impact of the spontaneous emission was not observed yet. Instead, the quality of CEP stabilization has often been found to be strongly influenced by technical noise of the pump laser [72, 123] which is converted via nonlinear interactions inside the cavity into a jitter of the per-roundtrip CEP shift $\Delta\varphi_{CE}$, i.e., in the CEF f_{CE}.

It is shown in this chapter that the reversal of the analysis of the CEP stabilization results and interpreting the CEF as power detector gives new insight into the power-to-CEF conversion inside a mode-locked laser. Surprisingly, it is found that this conversion step features a jitter of the CEF below the level that is expected for a transfer of pump laser shot noise into CEF jitter. Feedback squeezing [132–134], and the nonlinear interaction mechanisms of a soliton induced photon-number squeezing and a quantum non-demolition (QND) measurement are discussed as possible origin for this feature. Furthermore, it is found that, irregardless of the exact explanation, for each mode-locked laser there still exists a fundamental limitation to the achievable CEP jitter, which is considerably reduced compared to the shot-noise level. This limitation does, however, not constrain the accessible uncertainty in frequency metrology.

4.1. Observation of Sub-Shot-Noise Carrier-Envelope Frequency Jitter

The shot-noise limit of photodetection results from the fact that light can be absorbed by matter in discrete energy packets only [135], which is considered one of the foremost foundations of quantum optics. The quantum nature of light, in particular the statistical nature of temporal correlation between photons, has been carefully tested in numerous experiments [136–138].

These experiments revealed that coherent light sources emit temporally uncorrelated photons with a Poissonian photon number statistics. As a result, the variance σ for the detection of N photons from a laser is given by $\sigma = \sqrt{N}$. This leads to the standard shot-noise level, which is the relative uncertainty

$$\Delta P_{\mathrm{SNL}} = \frac{\sigma}{N} = \frac{1}{\sqrt{N}} \tag{4.1}$$

in the determination of an optical power $P = N E_{\mathrm{phot}}/t_{\mathrm{meas}}$ [139, 140] by counting the number of photons with energy $E_{\mathrm{phot}} = h\nu$ absorbed in a time span t_{meas}. Detection of light from other classical light sources, e.g., thermal light, has been found to exceed the standard shot-noise limit of a coherent source due to the differing photon number statistics. Thus, it is impossible to overcome the fundamental shot-noise limitation of an absorptive photodetector even when all technical noise contributions are eliminated in a classical light source.

Figure 4.1 sketches how a KLM laser acts as a power meter for the pump laser. The measured pump beam is coupled into the cavity of a mode-locked laser and creates an inversion inside the gain medium. The emitted light experiences self-amplitude modulation through aperturing and eventually laser pulses form out. Due to their high peak power, these pulses experience the Kerr nonlinearity of the gain medium. Under stationary conditions, the combined action of Kerr nonlinearity, self-amplitude modulation, and dispersion compensation generates pulses having the same shape after each roundtrip. However, the CEP is not reproduced after each roundtrip and shifts by an amount which is influenced by the intracavity power and, therefore, by the external pumping. Thus, the pump power can be measured via the CEP drift rate, which is measured on the fraction of the pulses that is coupled out. The energy coupled out of the cavity is subsequently replenished by laser gain, maintaining constancy of the intracavity power.

In the following experiments, a highly power-stabilized continuous-wave laser is utilized as pump source. This pump laser emits 4 W of optical power at 532 nm. Previous investigations reported relative intensity noise (RIN) of this source of a few 10^{-7} Hz$^{-1/2}$ in the 1 to 100 kHz range [72]. Despite the generally acknowledged quality of this light source, this value is still a factor $\approx 100 = +20$ dB above shot noise (estimated RIN $\sim 10^{-9}$ Hz$^{-1/2}$). In the following, the unit dBs will be used in

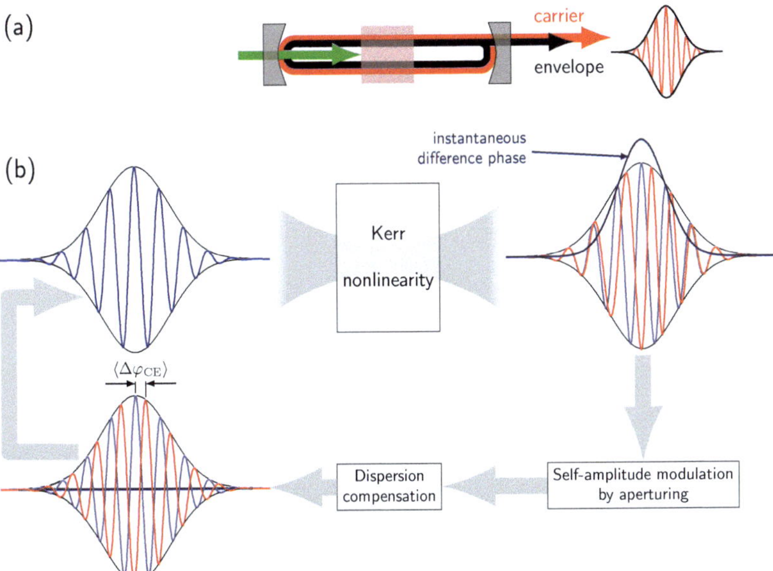

Figure 4.1.: (a) Scheme of using the mode-locked laser cavity as light power detector. (b) Illustration of the conversion of pump power to a shift of the carrier-envelope phase that happens inside an Kerr-lens mode-locked laser every roundtrip of the laser pulse. The Kerr nonlinearity experienced by the pulses in the gain medium leads to a modulation of the spectral phase and to self-focusing. By the action of aperturing and dispersion compensation stable pulse shapes form out, while the carrier-envelope phase is subject to a shift.

order to specify noise levels in decibel relative to the standard shot-noise level, i.e., the shot-noise enhancement or reduction factor R in dBs units is calculated by

$$R = 10 \log_{10} \left(\frac{\Delta P}{\Delta P_{\mathrm{SNL}}} \right) \tag{4.2}$$

(ΔP denotes the experimentally observed noise level in the determination of the optical power). The green light from this source is used to pump a femtosecond Ti:sapphire laser oscillator, operating at a center wavelength of 800 nm, repetition rate of 88 MHz, with a 10 fs pulse duration. During the pumping process 80% of the green pump light is absorbed inside the $L = 2.6$ mm long Ti:sapphire laser crystal and is converted into a train of femtosecond pulses circulating inside the cavity with 20% output coupling. For the pumping level of $P_{\mathrm{pump}} = 4$ W, pulses with an average power $P_{\mathrm{intra}} = 3$ W are generated at $\lambda = 800$ nm, i.e., the pump photon flow – which is the number of photons per second – is nearly identical to the intracavity photon flow

$$F = \frac{P_{\mathrm{intra}}\lambda}{hc} \approx 1.2 \times 10^{19} \ \mathrm{s}^{-1} \tag{4.3}$$

(h Planck constant, c velocity of light). This yields (single-sided) shot-noise levels of

$$\mathrm{RIN} = \sqrt{\frac{2}{F}} \approx 4.1 \times 10^{-10} \ \mathrm{Hz}^{-1/2} \, . \tag{4.4}$$

The frequency up to which the pumped laser directly follows changes in the pump power is estimated to be the gain relaxation oscillation frequency $f_{\mathrm{relax}} \sim 1$ MHz [70] for the experimental conditions used.

The shot-noise level of the intracavity field converts via the Kerr-interaction into a corresponding shot-noise-equivalent level of the CEF jitter. Under the absence of effects generating non-classical states of light inside the KLM laser's cavity, this is expected to be the lowest obtainable CEF jitter. It is important to note that, by the fact that the Kerr interaction inside the gain medium of the oscillator cavity takes place every roundtrip, CEP changes are converted into the measurement of variations of the CEF. With repetition rates f_{rep} on the order of 100 MHz, Eq. (1.8) predicts that even tiny phase changes of 100 nrad — corresponding to per-roundtrip timing differences of just a few ten yoctoseconds (10^{-24} s) — cause measurable frequency changes on the order of 1 Hz.

Characterization of the Sensitivity to Pump Power Modulations

As a first step towards characterization of the sensitivity of the KLM laser with respect to pump power fluctuations, the CEF response to pertubative ($< 0.17\%$)

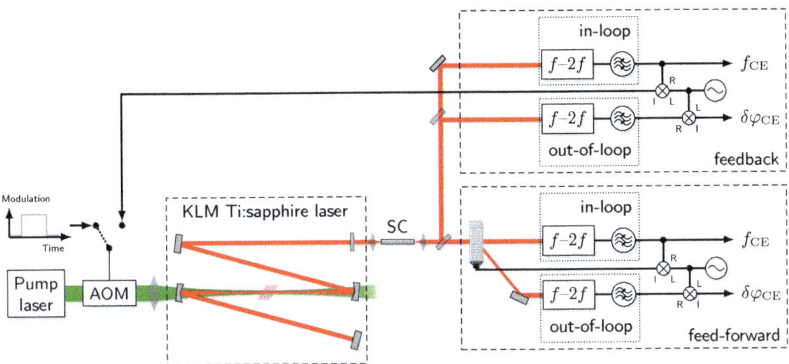

Figure 4.2.: Setup for the characterization of the sensitivity of the carrier-envelope phase $\delta\varphi_{CE}$ and carrier-envelope frequency f_{CE} to power fluctuations of the pump beam. (AOM: acousto-optic modulator; SC: supercontinuum generation; KLM: Kerr-lens mode-locked)

rectangular modulations of the oscillator pump power is evaluated using the setup shown in Fig. 4.2. To this end, an AOM is placed inside the path of the pump laser and driven with a rectangular signal, and the change of f_{CE} is monitored. Therefore, the beat signal is digitized with a storage oscilloscope and the recorded time series are numerically frequency-demodulated using the Takeda algorithm [126] to retrieve the temporal change $\Delta f_{CE}(t)$ in the CEF. The response with respect to the edge of the modulation transient is shown in Fig. 4.3. A careful analysis of $\Delta f_{CE}(t)$ reveals that the resulting transient exhibits a bi-exponential behavior with a dominant time constant of $\approx 1.5\ \mu$s and a secondary slower one at $\approx 65\ \mu$s with a relative weight of $\approx 15\%$. The latter slower process probably refers to thermally induced refraction changes in the Ti:sapphire crystal caused by absorption of the green pump light in the laser process (quantum defect $\approx 35\%$). The faster time constant, in contrast, stems from Kerr-type contributions, which can follow over the entire modulation bandwidth of the AOM and are otherwise only limited by the frequency f_{relax} of gain relaxation oscillations.

Furthermore, it is quite instructive to analyze fluctuations of the laser power in the 100 μs before the switching transient sets in, which is shown by the inset of Fig. 4.3. The RMS fluctuations of f_{CE} in this interval amount to only 1.1 kHz, which translates, by using the coupling factor determined below, into 40 μW power fluctuations on top of the 4 W pump laser beam. Considering the measurement bandwidth of ≈ 1 MHz and the averaging of 60 traces used in this measurement, these fluctuations translate into a RIN of 3×10^{-8} Hz$^{-1/2}$, i.e., a value of $+20$ dBs,

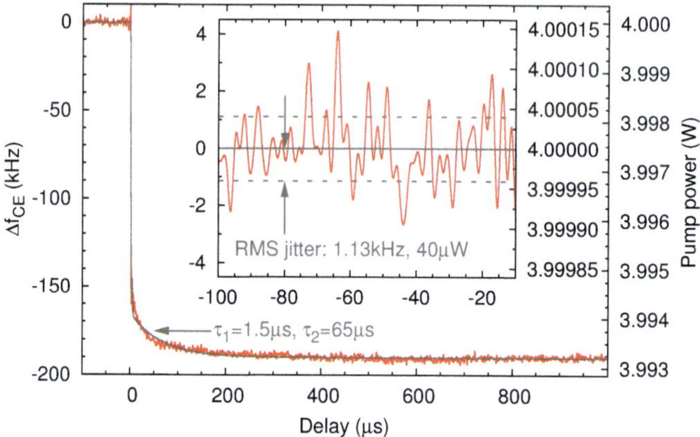

Figure 4.3.: Transient response of the carrier-envelope frequency to a $< 0.17\%$ decrease in the pump power of 4 W at delay $\Delta t = 0$. A bi-exponential fit to the data (gray) reveals a dominant fast contribution with a ≈ 1.5 µs time constant and a slower contribution with $\approx 15\%$ relative weight and a time constant of ≈ 65 µs.

which evidences the dominance of technical noise contributions for the pump laser used.

Figure 4.4(a) shows the frequency dependence of the coupling factor $\kappa = df_{CE}/dP_{pump}$ of pump power modulation dP_{pump} into CEF changes df_{CE}. This factor has been determined by synchronizing a frequency counter alternatingly to the time periods of strong and weak pump power modulation and extracting df_{CE} from these two measurement pairs. The measured coupling factors are almost constant at a level of 28 MHz/W for modulation frequencies up to 1 kHz. Towards higher frequencies there is a weak indication of a roll-off. Given that the time constant of thermal CEF variations is 65 µs, it is estimated that CEF modulation due to thermal refractive index changes can only follow up to a frequency of 15 kHz. Hence, this roll-off of the coupling factor at high frequencies probably stems from a stronger averaging of these contributions to the CEF variations.

Finally, the orange trace in Fig. 4.4(b) shows a spectral analysis of the pump laser RIN over the range from 2 Hz to 50 kHz. This curve has been obtained by applying the above mentioned numerical extraction of CEF variations on data collected for a free-running Ti:sapphire oscillator. These measurements directly reveal that RIN reaches values up to +40 dB above the shot-noise level (4.1) at frequencies of about 500 Hz. The excellent agreement with comparable previous measurements

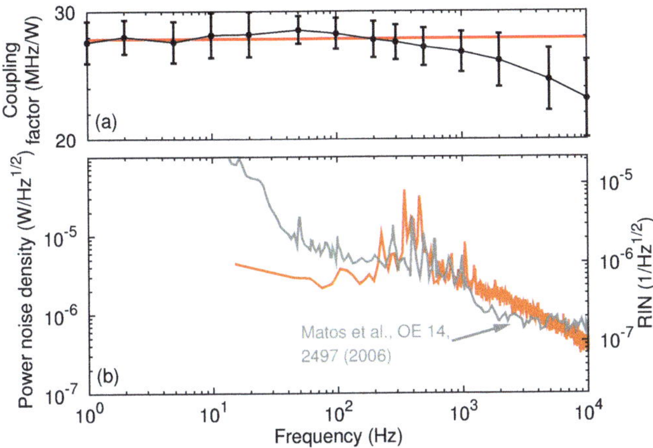

Figure 4.4.: Characterization of the spectral sensitivity of the Kerr-lens mode-locked laser as power detector. (a) Coupling factor of external pump power modulation into changes of the carrier-envelope frequency. This factor has been determined with a rectangular modulation of the pump power with a modulation depth of $< 0.17\%$ imposed by an acousto-optic modulator. Then the periodicity of the rectangle has been varied and the carrier-envelope frequency difference has been measured by synchronizing a frequency counter alternatingly to time periods of strong and weak pump power modulation by the acousto-optic modulator driver signal. For frequencies up to ≈ 1 kHz, the coupling factor is nearly constant at 28 MHz/W. (b) Power noise density of the pump laser (orange) as deduced from measured carrier-envelope frequency variations using the coupling factor from (a). Direct photo-absorptive measurement on a very similar laser shown for comparison (gray curve, extracted from [72]).

(gray) of the RIN employing a photo-absorptive detector on a similar pump laser [72] confirms that the CEF is a valid measure for the pump laser RIN. Furthermore, the agreement shows that in this frequency range the pump laser RIN dominates over other noise sources inside the KLM laser. Hence, estimation of the power detection sensitivity of the KLM laser requires suppression of the pump laser's power noise.

Sub-Shot-Noise CEF Variations in Feed-Forward CEP Stabilizations

In order to overcome this limitation, the CEP jitter in stabilized operation mode is analyzed in the following. Ideally, the CEP stabilization compensates for all technical RIN sources within the bandwidth of the stabilization and reveals the shot-noise-equivalent level of the CEF jitter within this bandwidth.

Figure 4.5 shows the power noise density inferred from the CEP jitter measurements employing the feed-forward CEP stabilization. The PND $S_\varphi(f)$ measured with the setup shown in Fig. 4.2 is converted to a power noise density via multiplication by the Fourier frequency f [141] and by normalizing the result using the calibration factor κ from Fig. 4.4(a)

$$S_P(f) = \frac{S_\varphi(f) \cdot f}{\kappa} \,. \tag{4.5}$$

Figure 4.6 illustrates that, due to the multiplication with f in Eq. (4.5), linearly decreasing parts of the PND curve translate into a constant level of the power noise densities, and constant parts of the PND result in a linear increase of the power noise density.

In fact, the power noise density inferred from the PND achieved with the feed-forward stabilization is not limited by shot noise and clearly reaches values below the shot-noise level in the low-frequency region between 100 and 500 Hz. In the complete interval between approximately 10 and 100 Hz, the stabilization results in Fig. 4.5 indicate optimum noise performance with RIN levels of

$$\mathrm{RIN}(f) = \frac{S_P(f)}{P_{\mathrm{pump}}} \approx 2 - 4 \times 10^{-11} \ \mathrm{Hz}^{-1/2} \,. \tag{4.6}$$

Such RIN levels correspond to pump power fluctuations on the order of 100 pW/Hz$^{1/2}$ on top of the 4 W pump beam. Recalling that the conventional shot-noise limit of photo-absorptive detection is 4.1×10^{-10} Hz$^{-1/2}$, these measurements yield shot-noise suppression factors below -10 dBs. In the optimum, these measurements indicate that noise levels of -13.4 dBs, i.e., down to 22 times below the standard shot-noise level, are achieved. The frequency range of noise suppression seems only to be limited by the high-frequency sensitivity roll-off that is caused by beat note detection noise. This detection noise of the f_{CE} results in an increase of the detected power noise densities which is linear in a double logarithmic representation

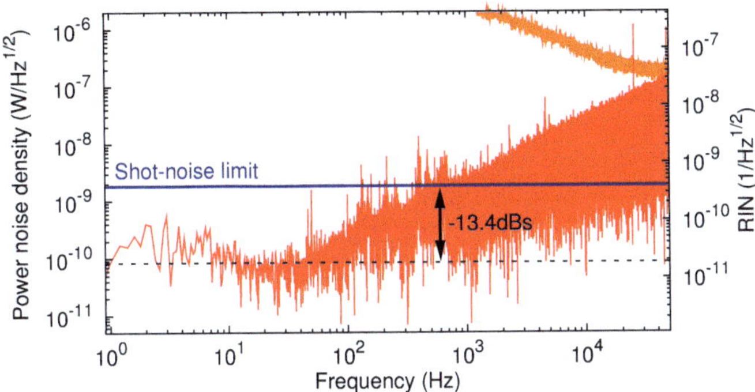

Figure 4.5.: Power noise detection limit (red) evaluated by the measurement setup shown in Fig. 4.2. Here, the feed-forward carrier-envelope phase stabilization is employed for elimination of technical noise contributions to the carrier-envelope frequency jitter. Single-sided shot noise density (blue) for a photon flow of 1.2×10^{19} s^{-1}. In the minimum around 30 Hz, the relative intensity noise (RIN) corresponding to the CEP fluctuation in the stabilized beam reaches values ≈ 13.4 dB below the shot-noise level. In addition the high-frequency tail of the free-running carrier-envelope phase noise from Fig. 4.4 is included as orange trace for illustration. The discrepancy between the progression of the orange and the red trace is caused by the carrier-envelope phase stabilization.

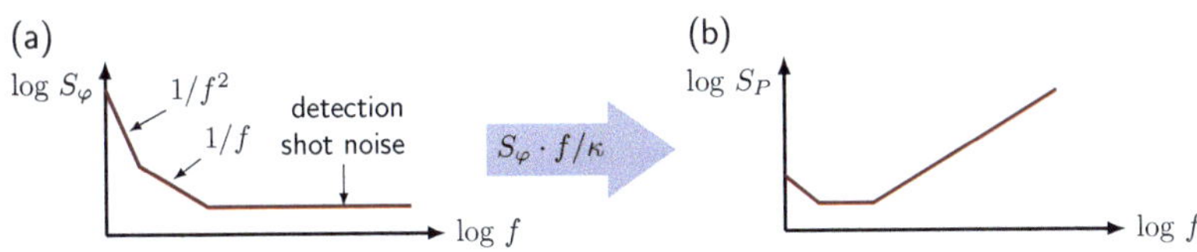

Figure 4.6.: Conversion of basic trends in phase noise density curves (a) to properties of the power noise densities (b) due to Eq. (4.5).

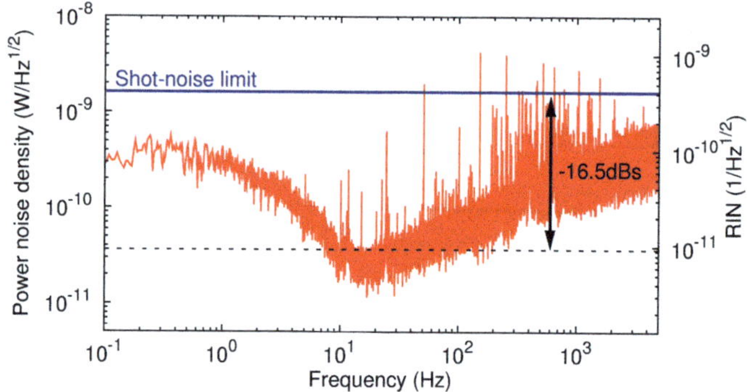

Figure 4.7.: Power noise detection limit (red) evaluated by measurements using the feedback CEP stabilization method. Hence, the acousto-optic modulator placed in the pump beam is used to suppress contributions from the pump laser's technical noise to the carrier-envelope frequency jitter. Single-sided shot noise density (blue) for a photon flow of $1.2 \times 10^{19}\,\mathrm{s}^{-1}$. In the minimum around 15 Hz, the relative intensity noise (RIN) corresponding to the residual carrier-envelope phase fluctuations reaches values of ≈ 16.5 dB below the shot-noise level.

(see discussion of Fig. 4.6) and deteriorates the sensitivity to power fluctuations by about one order of magnitude per decade frequency increase. On the low frequency side, a slight increase of power noise densities is observed, which may result from a relative drift of the two interferometers of the OOL characterization.

In the feed-forward stabilization discussed up to now, the stabilization mechanism acts on the result – the CEF jitter – of the conversion of pump power to CEP drift rate taking place inside the mode-locked laser. In the following paragraph, it is shown that the sub-shot-noise CEF jitter achieved by CEP stabilization is not specific to the feed-forward stabilization method, but also observable for the feedback based CEP stabilization.

Sub-Shot-Noise CEF Variations in Feedback Based CEP Stabilizations

For a further evidence of sub-shot-noise CEF statistics in CEP stabilized pulse trains the measurements are repeated using the servo loop approach for CEP stabilization. Since the previous measurement revealed that the sub-shot-noise CEF statistics are most clearly observable in the low frequency region, the measurement is focused on collecting more data for frequencies below 10 Hz. In order to render these measurement as free from spurious noise as possible, the interferometer topology is changed from the PQCP design used for the feed-forward results to the DQCP design studied in Section 2.1.2, and an analog PLL is employed. The beat note signal-to-noise ratios in this configuration are slightly higher and amount to $\approx$ 40 dB (RBW 100 kHz, VBW 300 kHz) in both IL and OOL interferometer.

As it is shown in Fig. 4.7, even for the feedback stabilization the CEF jitter clearly displays sub-shot-noise performance. Due to the factor of 10 increase in the signal-to-noise ratio of the beat note in the OOL interferometer, the power noise density levels in the roll-off region at frequencies above 10 Hz are lessened by about a factor of 10. As a result, the power noise densities are below the standard shot-noise over the complete measurement range from 0.1 Hz to 5 kHz. At the minimum at $\approx$ 15 Hz, noise densities of -16.5 dBs are achieved. Towards lower frequencies, the increase of the PND is resolved more clearly than in the feed-forward measurement. This feature corresponds to an increase of the PND that is steeper than a $1/f$-drift contribution. Interestingly, this contribution saturates towards lower frequencies.

Discussion

In order to gain further insight into the noise generating mechanisms, the measurement are further analyzed using Allan deviations. This time domain based measure is often utilized for the discrimination of different noise sources [141] and has its origin in the analysis of the frequency instability of time standards (see, e.g., Refs. [15, 107] for application to frequency combs). Typically, the modified Allan deviation is used due to its stronger capability to distinguish between different noise sources [141]. The modified Allan deviation of the frequency deviation $\Delta f = (2\pi)^{-1} d\delta\varphi_{\mathrm{CE}}/dt$ for averaging times $\tau = m\tau_0$ is given by the square root of the two-sample variance [142]

$$\mathrm{Mod}\,\sigma_{\Delta f}(\tau) = \sqrt{\frac{1}{2m^4\left(M - 3m + 2\right)} \sum_{j=1}^{M-3m+2}\left\{\sum_{i=j}^{j+m-1}\sum_{k=i}^{i+m-1}\left[\Delta f_{k+m} - \Delta f_k\right]\right\}^2}\,,$$

$$(4.7)$$

where τ_0 is the fundamental sampling time of the frequency deviation Δf and $M\tau_0$ is the total measurement period.

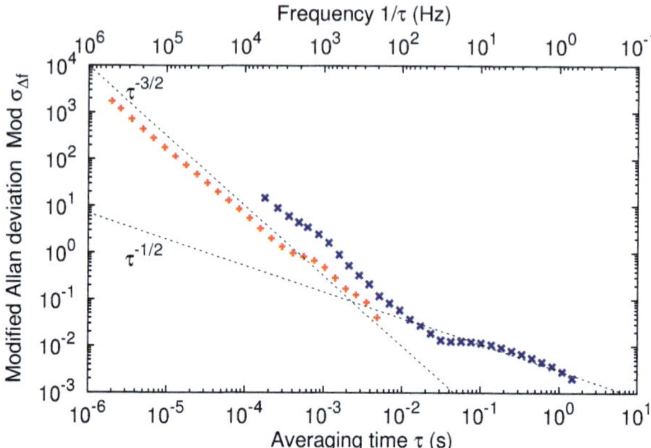

Figure 4.8.: Characterization of the frequency fluctuations Δf in the feed-forward (red) and feedback (blue) measurement using the Modified Allan deviation Mod $\sigma_{\Delta f}$.

The evolution of the modified Allan deviation Mod $\sigma_{\Delta f}$ over the averaging time τ is plotted in Fig. 4.8. Analysis of the curves shows that for the feed-forward measurement the modified Allan deviation decreases with a slope of ≈ 1.3 on a double logarithmic scale in the interval between 2 and 400 µs. This is indicative for the dominance of white phase noise, for which a decrease proportional to $\tau^{-3/2}$ is expected [141]. A similar slope is also observed for the feed-forward measurement between 2 and 30 ms and for the feedback trace between 200 µs and 10 ms. These observations are consistent with the finding that in the PND corresponding to the analyzed feed-forward and feedback measurement a detection noise induced constant level is observed with just a few distinct contributions exceeding this level. Furthermore, the modified Allan deviation of the feedback measurement clearly exhibits a change in slope for averaging times above 1 ms. For these longer times the curve approximately decreases as $\approx \tau^{-1/2}$, which is the slope for a process with white frequency noise. Hence, the modified Allan deviation indicates the presence of white frequency noise on long observation times in the feedback measurement. According to Eq. 4.5, such white frequency noise corresponds to white power noise such as the pump laser's shot noise. In addition, the modified Allan deviations show that, within the period analyzed by the measurements, longer averaging intervals leads to lower values of the modified Allan deviation, i.e., to lower uncertainties in the estimation of the frequency. Thus, the presence of such a shot-noise-induced white CEF jitter does not limit the precision achievable in frequency metrology measurements.

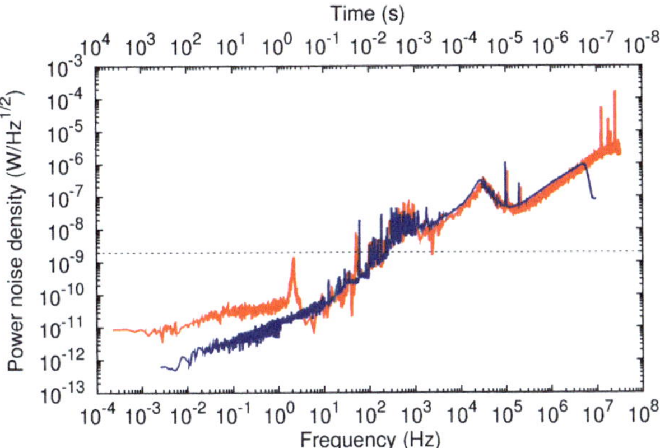

Figure 4.9.: Comparison with previous carrier-envelope phase stabilization results extracted form the Refs. [73] (red) and [74] (blue). For conversion of the phase noise density from Ref. [73], a coupling factor of 10 MHz measured on a similar Ti:Sapphire oscillator is used. For converting the results reported in Ref. [74], a coupling factor of 30 MHz derived in Ref. [96] by numerical simulations of the carrier-envelope phase shift in that specific laser is applied. The dashed black line shows the estimated shot-noise level of these measurements.

In addition, the observation of sub-shot-noise CEF jitter is not specific to the laser system and experimental setup that has been used in the above presented measurements. Such sub-shot-noise statistics are also observed in other feedback based CEP stabilizations, e.g., in the measurements of Fuji et al. [73] and Crespo et al. [74], which are plotted in Fig. 4.9. Although the coupling factors and the pumping level of the lasers used in these measurements are only approximately known (see discussion in the caption of Fig. 4.9), the plotted power noise densities show that sub-shot-noise CEF jitter is also achieved in these KLM lasers. Noise levels approximately 1000 times below shot noise, i.e., -30 dBs, are deduced, such that the finding of sub-shot-noise CEF jitter in these measurements is fulfilled even if the estimates are one order of magnitude off.

The setups of Fuji et al. and Crespo et al. differ from the setup used in above presented measurements in the way that no spectral broadening inside an MSF was required. While the measurement of Fuji et al. employed spectral broadening inside the PPLN generating the difference frequency for the f–0 scheme, the laser of Crespo et al. was already octave spanning such that the output could be directly used for f–$2f$ interferometry. This rules out that the sub-shot-noise CEF jitter is created by nonlinear effects in the MSF. Furthermore, the two mode-locked lasers differ in the repetition rate. While the laser of Crespo et al. has a significantly higher repetition rate of 500 MHz, the laser of Fuji et al. emitted pulses with 70 MHz. In both of these experiments beat note signal-to-noise ratios > 50 dB have been achieved resulting in a greater suppression of the beat note detection noise roll-off at high frequencies. Given the different laser designs, it is quite surprising that over the wide spectral range from 10 Hz to 5 MHz both measurements exhibit the same progression and features, which only seems explainable that great part of this noise is inherited from the pump laser, which was of the same model series and brand in both measurements. In contrast to the expectation of a pure conversion of shot noise, however, also in these measurements no constant noise floor resulting from a statistical CEF jitter is observed in the power noise densities. Only in the measurement of Fuji et al. [73] a slight leveling off of the power noise density is observed for frequencies below 1 Hz.

Furthermore, due to the differences of the applied stabilization schemes more advanced conclusions can be drawn. In the feed-forward stabilization, the actuator is placed behind the KLM laser. Hence, the feed-forward stabilization only acts on the classical result of the pump power-to-CEF conversion happening inside the laser. Therefore, this stabilization scheme is not limited by the quantum limit given by shot noise, and, analogous to the generation of squeezed light by using a feed-forward intensity stabilization in combination with the correlation in twin beams [143], lower noise levels are accessible. Therefore, the remaining task is the clarification of the mechanism generating the shot-noise suppression in feedback configuration, which has to include the conversion of pump power to CEF inside the KLM laser.

4.2. Origin of Sub-Shot-Noise Carrier-Envelope Frequency Jitter in Feedback Configuration

Measurements with a sensitivity below the standard shot-noise level have already been achieved in other optical setups. These measurements have been realized by either manipulating the quantum statistics of light or by exploiting a nonlinear optical interaction. To the end of identifying the origin of the sub-shot-noise CEF jitter, the feedback stabilization mechanism and the conversion of intracavity power to CEP drift rate inside a mode-locked laser is analyzed in more detail in the following sections.

4.2.1. Feedback Based Generation of Sub-Shot-Noise Characteristics

In the approach to noise cancellation, which is known as optical squeezing [144, 145], correlations between photons are actively enforced for generating non-classical quantum statistics of light. Depending on the method employed, different types of squeezing are achieved. By using, for example, the often employed technique of optical parametric frequency conversion the fluctuations in one of the two quadratures of the electric field may be reduced at the expense of increased fluctuations in the other quadrature [145]. This is different from squeezing of the fluctuations of the electric field's amplitude by making use of, e.g., the nonlinear propagation of a soliton in an optical fiber [146, 147].

While the above mentioned techniques are based on nonlinear optical processes, sub-shot-noise power spectra can also be generated in feedback loops employing linear optic modulators. This fact was observed for the first time in experiments with semiconductor laser diodes whose driving current was regulate via a feedback loop such that the output power of the laser diode is stabilized [132]. It was shown that the photocurrent fluctuation spectrum measured with a standard photodiode in negative feedback configuration can be decreased below the shot-noise level [132]. This is enabled by the fact that the laser diode is able to emit photon-number squeezed light if driven with a high-impedance constant-current source [148], which itself bases on the transfer the pump electron statistics onto the emitted photons due to a self-stabilization mechanism in the constant-current mode [149].

The theoretical analysis [134] of this feedback based photon-number squeezing reveals that the reason for the observation of reduced noise levels results from the manipulation of the photon statistics within the feedback loop [133, 134]. In contrast to quadrature squeezing, the magnitude of the uncertainty domain is not preserved by the action of the feedback. In addition, the analysis shows that the noise suppression enhances with increasing feedback gain such that the suppression is only limited by the stability regime of the feedback loop. Unfortunately, light beams with reduced photon noise can not be extracted from the feedback loop [132, 150]. This is due to the anti-correlation of the incident light with the vacuum input at the beamsplitter which taps off the extracted beam [134].

The feedback based CEP stabilization scheme can be interpreted in the light of these previous experiments with semiconductor lasers. In the CEP stabilization, the pump laser of the femtosecond laser substitutes the driving current and the photodiode is replaced by the intracavity process of converting pump power to CEP shift in combination with the f–$2f$ interferometer. The great sensitivity of the KLM laser to pump power modulation increases the gain of the feedback loop and, hence, the achievable shot-noise suppression.

Whether it is the feedback mechanism which is responsible for the sub-shot-noise power spectra in the feedback CEP stabilization requires incorporation of the details of the intracavity conversion of power to CEP drift into the theoretical models for the semiconductor laser diode experiments. This is beyond the scope of the thesis. However, the nonlinearity of the intracavity conversion of power to CEP drift permits two alternative interpretations of the feedback results, which are elucidated in the following two sections.

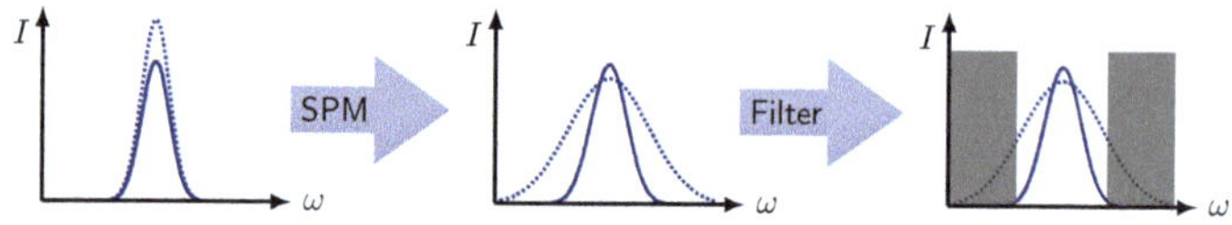

Figure 4.10.: Illustration of the mechanism leading to a photon-number squeezing by the combination of self-phase modulation (SPM) and spectral filtering. Soliton pulses with an intensity above the average value experience strong spectral broadening by SPM. Hence, filtering of the broadened spectrum in a narrow spectral filter leads to an increased loss for high amplitude pulses. Similarly, if the spectral filter is so narrow that it also introduces a loss for average pulses, pulses having an amplitude slightly below average are subject to higher transmission. Thus, the combination of SPM and spectral filtering may counteract amplitude fluctuations if the width of the spectral filter is chosen appropriately.

4.2.2. Interpretation in Terms of Soliton Induced Photon-Number Squeezing

In principle, the low levels of the CEF jitter observed in the experiment may also result from sub-shot-noise photon-number fluctuations of the cavity internal electric field generating the per-roundtrip CEP shift. Since KLM lasers are operated in the net anomalous dispersion regime and soliton-like pulse shaping is present, the mechanism underlying the sub-shot-noise CEF jitter may be comparable to the photon-number squeezing observed for two counterpropagating solitons in a fiber-optic Sagnac loop [147] or to the filtering induced soliton photon-number squeezing achieved for a single soliton transition through a fiber [146].

Due to the absence of counterpropagating pulses in the KLM laser cavity, only the photon-number squeezing by filtering comes into consideration for the explanation of sub-shot-noise CEF jitter. In order to observe the soliton induced photon-number squeezing, Friberg et al. launched picosecond pulses as solitons of a low order $N_{\mathrm{soliton}} > 1$ into a 1.5 km long optical fiber with anomalous group velocity dispersion. Due to the soliton order greater than one, these pulses experienced spectral broadening. By spectral filtering of the fiber output, they found that for certain spectral filtering bandwidths the integrated noise levels reach values up to 2.3 dB below the shot-noise level. This observation was heuristically explained by a nonlinear loss modulation mechanism that cancels out the photon-number fluctuations [146], as illustrated in Fig. 4.10. In a later publication, the authors showed that by launching solitons with a higher soliton order $N_{\mathrm{soliton}} \approx 3$ the temporal photon-number fluctuations with frequencies in the MHz range could be significantly reduced by 23 dB [151].

Transferred to contemporary dispersion-managed KLM lasers, these results indicate that the limited gain bandwidth of Ti:sapphire and the reflectivity characteristics of the cavity mirrors may provide suitable characteristics for soliton induced photon-number squeezing. Similarly, also the spatial filtering in the Kerr-lens induced self-amplitude modulation process may lead to such a self-stabilizing filtering action. In KLM lasers, spatial filtering typically only results in stabilization of the solitary pulse form against growth of noise in between the pulses and does not influence the pulse shaping process. Hence, the filtering in the spectral domain is regarded to have a stronger impact on possible photon-number squeezing. For the parameters of dispersion-managed KLM lasers, pulse shapes form out that are close to the soliton solution, i.e., $N_{\text{soliton}} \approx 1$, so that these lasers are not in the regime for soliton induced photon-number squeezing. Furthermore, if soliton squeezing plays such a strong role, one would also expect to see a significant reduction of technical noise contributions from the pump laser, e.g., in the free-running CEF jitter. However, this is not the case. Therefore, such a noise reduction mechanism is expected to play only a minor role for typical dispersion-managed KLM lasers.

4.2.3. Interpretation in Terms of Quantum Non-Demolition Measurements

As an alternative to the two above mentioned explanations, the sub-shot-noise CEF jitter can also be interpreted as the outcome of a QND measurement process. Before the actual interpretation is made and the connection to demonstrated QND measurements schemes is drawn, an introduction to QND measurements and the formalism that is used later on is given in the following paragraphs.

Introduction to Quantum Non-Demolition Measurements

In contrast to classical physics, quantum mechanics teaches us that every measurement disturbs the quantum mechanical system under study.[1] For measuring the eigenvalue of the observable $\hat{A}$ of interest, an interaction between quantum system and measurement apparatus has to be established. Thus, even when limited to a short period of time, there exists a contribution of the interaction to the Hamiltonian describing the temporal evolution of the quantum system. As a result, the system state after the measurement is different from the state just before the measurement.

---

[1]Note, this statement is based on the viewpoint that one knows about the exact quantum system state without measuring it. This standpoint is somewhat hypothetical since such information can only be obtained by measurements. Hence, a correct description of the temporal evolution must also account for the measurement apparatus quantum-mechanically [152]. Nevertheless, the following discussion adopts the simpler viewpoint that a measurement disturbs the quantum system.

The effect of this disturbance on the result of a measurement of a different observable $\hat{B}$ is determined by Heisenberg's uncertainty principle

$$\Delta\hat{A}\,\Delta\hat{B} \geq \frac{1}{2}|\langle[\hat{A},\hat{B}]_-\rangle|\,, \tag{4.8}$$

where the commutator

$$[\hat{A},\hat{B}]_- = \hat{A}\hat{B} - \hat{B}\hat{A}\,, \tag{4.9}$$

$\langle\cdot\rangle$ denotes the expectation value and

$$\Delta\hat{A} = \left[\left\langle(\hat{A}-\langle\hat{A}\rangle)^2\right\rangle\right]^{1/2} \tag{4.10}$$

is the variance (RMS fluctuation) of the measurement ($\Delta\hat{B}$ accordingly). If the two observables do not commute with each other

$$[\hat{A},\hat{B}]_- \neq 0\,, \tag{4.11}$$

the measurement uncertainty in $\hat{B}$ is the larger the higher the precision of the $\hat{A}$ measurement is. In addition to this direct contamination, the uncertainty in $\hat{B}$ may also act back to $\hat{A}$ if the subsequent free propagation of the system couples the two observables. Then, any subsequent measurement of $\hat{A}$ will exhibit a higher variance.

However, there are also quantum systems and observables for which such contamination feedback is not present. Such measurements are termed to be *quantum non-demolition* [153] or *back-action-evading* [154]. Theoretically, this non-demolition property allows for measurements with arbitrarily high accuracy and for the predictability of the outcome of measurements at every subsequent point in time.

Such QND measurements are most easily implemented indirectly by coupling the observable $\hat{A}$ of the quantum system $\mathcal{S}$ under study to some property of the so-called meter quantum system $\mathcal{M}$. This coupling introduces correlations between $\mathcal{S}$ and $\mathcal{M}$ which allow, after an appropriate "separation" of the two subsystems, for the determination of the variable $\hat{A}$ in the current state of $\mathcal{S}$ by inspection of $\mathcal{M}$ [155]. The key point here is the separation of the subsystems in such a way that the destructive detection of $\mathcal{M}$'s state does not lead to a contamination of $\mathcal{S}$.

Criteria for QND Measurements

In order to check whether an indirect measurement of a certain variable $\hat{A}$ fulfills the QND property, Caves et al. [154] developed the following two simultaneous conditions [156]:

1. Observable $\hat{A}$ is a QND variable if it is not affected by the free propagation of system $\mathcal{S}$, i.e., if $\hat{A}$ commutes with the unperturbed Hamiltonian $\hat{H}_0$ of $\mathcal{S}$ alone

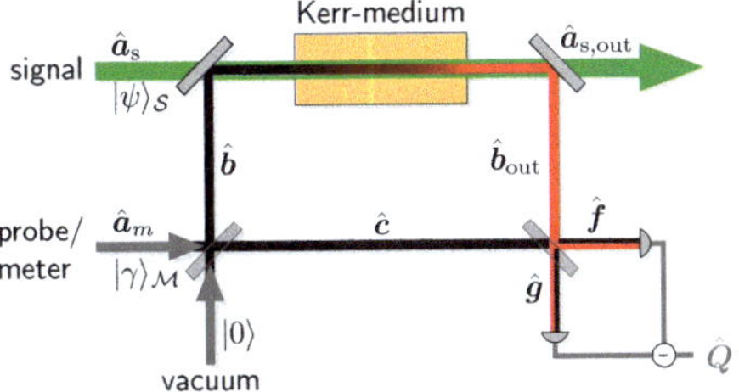

Figure 4.11.: Setup for quantum non-demolition measurement of the photon number proposed by Imoto et al. [157], which is based on detecting the nonlinear phase shift of the probe beam inside the Kerr-medium. The mirrors in the top row are assumed to be ideally dichroic such that they are transparent to the signal beam but reflective for the meter beam. The operator names written next to the different beams of the interferometer define the nomenclature and denote the corresponding annihilation operators.

$$[\hat{\boldsymbol{H}}_0, \hat{\boldsymbol{A}}]_- = 0\,. \tag{4.12}$$

This condition ensures that no back-action is encountered.

2. Measurement of $\hat{\boldsymbol{A}}$ is a QND measurement if it is unbiased by the interaction of $\mathcal{S}$ and $\mathcal{M}$

$$[\hat{\boldsymbol{H}}_I, \hat{\boldsymbol{A}}]_- = 0\,, \tag{4.13}$$

where $\hat{\boldsymbol{H}}_I$ is the interaction Hamiltonian between the two quantum systems.

QND Measurements in Optics

While the concepts of QND measurements were originally discussed in the context of detecting gravitational waves by the induced tiny displacement of Weber bars [158], optics has proven to be a fruitful domain to implement QND measurements [155].

Figure 4.11 depicts an optical QND scheme that has attracted a great deal of attention and which will be called Haus-Yamamoto interferometer (HYI) in the following [157]. This scheme serves to measure the light power of a signal wave, and is based on the transfer of information about the amplitude of a strong signal beam into the phase of a weak probe beam inside a Kerr-medium. The phase change is read out via embedding the Kerr-medium in one arm of the probe beam's Mach-Zehnder interferometer and homodyne balanced detection of the interferometer's two output ports.

As detailed below, the conversion of pump power in CEP drift rate can be interpreted in strong relationship to the transfer of information in the HYI. Hence, it is worth analyzing here in more detail along the lines of Refs. [157, 159]. The state of the joint quantum system consisting of signal and meter is denoted by

$$|\psi, \gamma\rangle = |\psi\rangle_{\mathcal{S}} \otimes |\gamma\rangle_{\mathcal{M}}, \tag{4.14}$$

and the incident meter beam is assumed to be in a coherent or Glauber state. Thus, it may be expanded in terms of the number states $|n\rangle$ as [160]

$$|\gamma\rangle_{\mathcal{M}} = e^{-|\gamma|^2/2} \sum_{n=0}^{\infty} \frac{\gamma^n}{\sqrt{n!}} |n\rangle. \tag{4.15}$$

The number states $|n\rangle$ are eigenstates of the photon number operator $\hat{N} = \hat{a}^\dagger \hat{a}$, where $\hat{a}^\dagger$ and $\hat{a}$ are the photon creation and annihilation operator, respectively (see [161] for detail). This assumption is valid for light produced by a laser pumped far above threshold, for which the quantum state of light can be described by a sequence of these coherent states with changing phases in adjacent sequences [162]. The first beamsplitter in Fig. 4.11 splits the meter beam into two parts. Due to the coherent input states, the reflected and transmitted beams are in coherent states $|i\sigma\gamma\rangle$ and $|\sqrt{1-\sigma^2}\gamma\rangle$, respectively. The reflected beam interacts with the signal beam inside a Kerr-medium. At the second beamsplitter, this beam then interferes again with the transmitted arm, which now serves as the local oscillator for the homodyne balanced detection.

Restricting the discussion to an exclusive cross-phase modulation, the interaction of meter and signal beam inside the Kerr-medium can be described via the Hamiltonian [159]

$$\hat{H}_I = \hbar K \hat{a}_s^\dagger \hat{a}_s \hat{b}^\dagger \hat{b}, \tag{4.16}$$

where K is a proportionality constant. The number of photons $\hat{b}^\dagger \hat{b}$ and $\hat{a}_s^\dagger \hat{a}_s$ in meter and signal beam, respectively, are constants of motion since no loss is introduced by the mutual interaction. Thus the equations of motion

$$\frac{d}{dt}\hat{b} = [\hat{H}_I, \hat{b}]_- = iK \hat{a}_s^\dagger \hat{a}_s \hat{b}, \tag{4.17}$$

(for $\hat{a}_\mathrm{s}$ equivalently) can be integrated directly to yield

$$\hat{b}_\mathrm{out} = e^{iKT_K \hat{a}_\mathrm{s}^\dagger \hat{a}_\mathrm{s}}\hat{b}, \qquad \hat{a}_\mathrm{s,out} = e^{iKT_K \hat{b}^\dagger \hat{b}}\hat{a}_\mathrm{s}, \tag{4.18}$$

where T_K denotes the interaction time given by the traveling time through the Kerr-medium. As a result, the charge operator of the balanced detector reads

$$\hat{Q} = -q\left(\hat{b}^\dagger \hat{c}\, e^{-i\hat{\Phi}_m} + e^{i\hat{\Phi}_m}\hat{c}^\dagger \hat{b}\right), \tag{4.19}$$

where q is the detector efficiency, $\hat{\Phi}_\mathrm{m} = KT_K\hat{a}_\mathrm{s}^\dagger\hat{a}_\mathrm{s}$ is the phase shift operator of the meter beam, and $\hat{c}^\dagger$ and $\hat{c}$ are the creation and annihilation operator of the transmitted beam. The expectation value of $\hat{Q}$

$$\langle \hat{Q}\rangle = \langle \psi, \gamma|\hat{Q}|\psi, \gamma\rangle \approx 2q\sigma\sqrt{1-\sigma^2}|\beta|^2 KT_K n_\mathrm{s} \tag{4.20}$$

shows that the measured signal is to first approximation linearly dependent on the signal photon number n_s. The smallest resolvable change in photon numbers and the distortion of the signal phase $\hat{\Phi}_\mathrm{s} = KT_K\hat{b}^\dagger\hat{b}$ associated with it are given by [152, 159]

$$\Delta n_\mathrm{s} \geq \frac{1}{2\sigma\sqrt{1-\sigma^2}|\gamma|KT_K} \tag{4.21}$$

$$\Delta\hat{\Phi}_\mathrm{s} = KT_K\sigma|\gamma|, \tag{4.22}$$

which shows that the mean square fluctuations fulfill the equality in the Heisenberg uncertainty principle (4.8) in the limit $\sigma \to 0$. As Eq. (4.21) shows, there is no fundamental limit for the uncertainty about the power of the signal wave, since theoretically the nonlinearity KT_K can be made arbitrarily large by increasing the interaction time T_K between signal and meter beam or by enhancing of the nonlinear coupling K using media offering a higher cross-phase modulation.

As the interaction inside the Kerr-medium is assumed to be of cross-phase type exclusively, the unperturbed Hamiltonian $\hat{H}_0$ in Eq. (4.12) is the identity for the HYI. Thus

$$[\hat{H}_0, \hat{N}_\mathrm{s}]_- = 0, \tag{4.23}$$

where $\hat{N}_\mathrm{s} = \hat{a}_s^\dagger \hat{a}_s$ is the indirectly measured signal beam photon number operator. Furthermore, $\hat{N}_\mathrm{s}$ also commutes with $\hat{H}_I$

$$[\hat{H}_I, \hat{N}_\mathrm{s}]_- = 0, \tag{4.24}$$

such that both QND criteria (4.12) and (4.13) are met for the HYI.

Effectiveness of QND Measurements

As any experimental implementation of a QND measurement will suffer from, e.g., the presence of loss, conditions (4.12) and (4.13) can only be fulfilled approximately. As a result, quantitative criteria are necessary for the objective evaluation of the effectiveness of the QND measurement scheme.

Following Ref. [155], a HYI measurement is of QND type if the following two criteria are met:

1. $T_s + T_m > 1$. Here T_m and T_s are the transfer coefficients of the signal-to-shot-noise ratios from the measured signal observable to the meter observable and vice versa. For the ideal HYI there are no losses, and $T_s = 1$. Using the so-called nonlinear cross-phase gain $g = 2\sqrt{\langle\hat{\Phi}_s\rangle\langle\hat{\Phi}_m\rangle}$ the transfer coefficient T_m can be written as

$$T_m = \frac{g^2}{1 + g^2}\,. \tag{4.25}$$

2. $V_{s|m} < 1$. Here $V_{s|m}$ is the conditional variance relative to shot noise, which quantifies the certainty about the signal beam's state given a specific result for a measurement of the meter's phase. Using the cross-phase gain g this condition reads

$$V_{s|m} = \frac{1}{1 + g^2} < 1\,. \tag{4.26}$$

The more clearly these two inequalities are fulfilled, the stronger is the QND-like behavior of the measurement. As the values of both quantities $T_s + T_m$ and $V_{s|m}$ are bounded to the interval $[0, 2]$, an optimal QND measurement is characterized by: $T_s + T_m = 2$ and $V_{s|m} = 0$.

Demonstrated Shot-Noise Suppression Results of Optical QND Measurements

While the concept of the HYI is structurally convincing and implementation seems to be straightforward, experimentalists had a hard time in really demonstrating sub-shot-noise performance. The main experimental difficulty is to find an appropriate nonlinear medium offering a strong cross-phase-modulation while not being absorptive or adding additional noise [155]. As a result, there are only two realization of the original HYI described in literature: while the first is based on utilization of collisions between optical fiber solitons [163, 164], the other utilizes atomic vapors embedded in an optical cavity [165–167] as nonlinear medium. Although both demonstration make use of intensification of the nonlinear coupling by increasing the interaction length or by enhancing the electric field in a cavity, the obtained sensitivity improvement has only been as low as -0.5 dBs [164] and -0.6 dBs [167], respectively. Only later, Roch et al. were able to further push the sensitivity of the atomic vapor based approach to -3.5 dBs [168], which is currently the best reported value for an optical QND measurement.

Apart from QND measurements in a HYI geometry, QND measurements of the electric field's quadrature amplitudes have been shown by four-wave mixing in optical fibers [169] with nearly no sensitivity improvement but also no enhancement of noise. Furthermore, parametric processes have been exploited for QND measurements: after the first paper of La Porta et al. demonstrating -0.6 dBs [170], implementation of the parametric process in a non-collinear optical parametric amplifier allowed pushing the sensitivity to -1.5 dBs [171]. By making use of squeezed input to the parametric process, the sensitivity has been raised further to -2.8 dBs [172]. Employing parametric processes, repeated QND measurements have been demonstrated for the first time and conditional variances of -1.8 dB [173] and -2.9 dB [174] between the two QND measurements have been obtained.

The Pump Power to CEP Drift Conversion as QND Measurement

The conversion of power to per-roundtrip CEP shift can be interpreted as a special case of a QND measurement: while the pulse properties may change during one roundtrip, the mode-locking mechanism enforces generation of pulses with identical pulse energy, spectrum, and spectral phase after each roundtrip. Within one roundtrip, a certain amount of quantum noise is added to the pulse by output coupling and amplification. The CEP, however, is shifted, and the amount of this shift depends on the intracavity peak power, as discussed in Section 1.6. Thus, when looking at the pulses stroboscopically after each roundtrip, the intracavity peak power is measured by the shift in the CEP while leaving all other pulse parameters in the same state. In consequence, analyzing the CEP drift in such a configuration can be viewed as a stroboscopic QND measurement of the intracavity peak power, with the CEP being the conjugate variable which the information is transfered to. In this analogy, the envelope of the pulses takes over the part of the meter beam in the HYI, such that the envelope's maximum defines the phase reference. Furthermore, since the intracavity peak power follows the external pump within the cavity bandwidth [70, 94] the CEP drift essentially measures the power of the pumping laser beam.

Compared to light power measurements using the HYI, the power-to-CEF conversion exploits intracavity femtosecond pulses with a considerably larger peak power of $\hat{P} \approx 3$ MW. With above determined coupling factor of $\kappa = 28$ MHz/W the variation of the per-roundtrip CEP shift $\Delta\varphi_{\mathrm{CE}}$ with power is calculated to be:

$$\frac{d\Delta\varphi_{\mathrm{CE}}}{dP_{\mathrm{pump}}} = \frac{2\pi}{f_{\mathrm{rep}}}\frac{df_{\mathrm{CE}}}{dP_{\mathrm{pump}}} \approx 2\ \mathrm{rad/W}\,. \tag{4.27}$$

Hence, the nonlinear phase shift per roundtrip is

$$\varphi_{\mathrm{nl}} = P_{\mathrm{pump}}\frac{d\Delta\varphi_{\mathrm{CE}}}{dP_{\mathrm{pump}}} \approx 8\ \mathrm{rad}\,. \tag{4.28}$$

This agrees favorably with the per-roundtrip phase shift

$$\varphi_{\mathrm{nl}}^{\mathrm{soliton}} = \frac{L n_2 P_{\mathrm{intra}}}{2\lambda_c w_0^2 f_{\mathrm{rep}}\tau_{\mathrm{sech}}} \approx 4.6\ \mathrm{rad}\,, \tag{4.29}$$

derived from soliton perturbation theory, given that soliton perturbation theory only includes SPM as nonlinear interaction and deviations for dispersion-managed Ti:sapphire lasers generating ultrashort pulses are found to be as large as a factor of 2.6 [96] ($w_0 \approx 8\ \mu\mathrm{m}$ beam waist, $\tau_{\mathrm{sech}} = \tau_{\mathrm{FWHM}}/1.76$, $\tau_{\mathrm{FWHM}} \approx 10\ \mathrm{fs}$ pulse duration, $n_2 = 3 \times 10^{-20}\ \mathrm{m}^2/\mathrm{W}$ nonlinear index of refraction, $\lambda_c = 800\ \mathrm{nm}$ carrier wavelength).

Using the formalism for evaluating the effectiveness of QND measurements [155], the nonlinear coupling coefficient of $g = 2\varphi_{\mathrm{nl}}$ gives an estimate for the conditional variance of

$$V_{\mathrm{s|m}} = \frac{1}{1+g^2} \approx 0.004\,, \tag{4.30}$$

which specifies the shot-noise reduction of the QND measurement referenced to the standard shot-noise level. Hence, Eq. (4.30) predicts that for a typical femtosecond Ti:sapphire oscillator, such as the one used, already -24 dB sub-shot-noise performance is achievable. This is further supported by the soliton perturbation theory based estimate yielding a shot noise reduction of 20 dB.

These estimates come very close to the experimentally observed shot-noise reduction factor and deviate by only $4-8$ dB, which is readily explainable by the impact of uncompensated technical noise sources that inhibit reaching the theoretical QND shot-noise reduction. Surprisingly, however, the previous feedback based CEP jitter measurements by Crespo et al. shown in Fig. 4.9 demonstrate a higher shot-noise suppression of ≈ 30 dB at low frequencies for a laser that theoretical supports shot-noise suppression of ≈ 14 dB.[2] This indicates that there may be an additional

---

[2] The Ti:sapphire laser of Crespo et al. has nearly the same coupling factor of 30 MHz/W, a significantly higher repetition rate of 500 MHz, and was pumped with 6.2 W. Hence, the per-roundtrip CEP shift is $\Delta\varphi_{\mathrm{CE}} \approx 2.2$ rad. This results in a conditional variance of $V_{\mathrm{s|m}} \approx 0.044 = -13.6$ dB.

contribution of the feedback based squeezing of quantum fluctuations, as discussed above.

4.3. Concluding Remarks

It has been shown that the CEF of a feed-forward and feedback CEP-stabilized mode-locked laser exhibits fluctuations which are below the levels expected for a conversion of pump laser shot noise into CEF jitter. In search for an explanation for these – especially for the case of the feedback CEP stabilization – surprising results, a feedback based suppression of noise levels and a QND measurement have been identified as two possible origins. In case of the feedback mechanism the nonlinear backaction serves to manipulate fluctuations of the intracavity field which convert into the CEF jitter. In contrast, the QND measurement bases on the nonlinear optical interaction which is involved in this conversion. While, as a result of the mode-locking mechanism, properties like spectrum, spectral phase and power of the circulating pulse are restored after each roundtrip, the CEP is shifted due to nonlinear propagation inside the gain medium by an amount that is related to the pump power. Thus, in this QND measurement the CEP is the conjugate variable, which carries the information of interest, such that the pulse envelope serves as phase reference instead of the reference arm in the HYI. As a consequence, the pulse is signal and meter beam simultaneously, which makes the QND measurement process intrinsically collinear and free of spurious noise contributions of the HYI stemming from interferometer arm length jitter or beam pointing variations. Whether the feedback mechanism or the QND measurement, or a combination of both is responsible for the observed sub-shot-noise CEF jitter cannot be answered with the currently available data. The experimentally obtained shot-noise suppression values of up to -16.5 dBs are consistent with both interpretations. For the QND interpretation the conditional variance gives estimates which predict 8 dB stronger noise suppression than observed experimentally. For the feedback squeezing mechanism shot-noise suppression values are only limited by the stability of the feedback loop. Estimation of the feedback based shot-noise suppression values requires the development of a quantum-mechanical model for the feedback loop including the nonlinear conversion of intracavity power fluctuation into CEF jitter, which is beyond the scope of this thesis. Irregardless of the fact which of the two explanations is the valid one or if a combination of both is responsible for the sub-shot-noise feature, the quantum noise is the source for the ultimate limit for carrier-envelope stabilization and only the dynamics (stability of the feedback loop) or properties (nonlinear phase shift) of the mode-locked laser under study determine how strong it can be suppressed.

If future experiments prove the QND process to be the dominant contribution, this will have dramatic implications. Then the CEF jitter of a KLM laser can be used as a detector for faint power modulations, which is not only remarkable in its sensitivity but also due to the frequency range: demonstrations of squeezing and QND measurements have often relied on shot-noise-limited laser radiation as noise reference. This is obtained in frequency regions well above f_relax, i.e., at MHz frequencies, as in this frequency range technical noise contributions are absent. However, if the CEF jitter serves as light power measurement scheme, sub-shot-noise performance is achieved in the low frequency region between 1 and 400 Hz, which makes it ideally suited for application in gravitational wave interferometers. Often these specially designed interferometers reach their largest sensitivities at a few 100 Hz [175, 176]. For frequencies above $\approx$ 100 Hz, the performance of such an CEF jitter based power detector slowly rolls off due to the significant influence of detection noise of the beat note causing a deterioration by about one order of magnitude per decade frequency increase. Therefore, boosting the detection efficiency, e.g., by optically amplifying the signal in the $2f$ arm prior to frequency-doubling may serve to push this roll-off towards higher frequencies.

In both interpretations, the great magnitude of shot-noise suppression mainly results from the integration of the Kerr effect into a laser cavity. In the one interpretation, this integration enables the mode-locking mechanism to influence the power-to-CEF conversion such that it resembles a QND measurement. In the other interpretation, this integration makes the CEF very sensitive to pump power modulations and, hence, increases the gain of the feedback loop. Utilization of an f–$2f$ interferometer to resolve the experienced phase shift effectively converts the phase measurement into a frequency measurement and enables the use of the fully developed methodology of frequency metrology [13, 177]. As frequencies are ultimately the quantities that are most precisely determinable, conversion of power fluctuations into frequency fluctuations, therefore, greatly improves the detectivity of the probed physical entity. Similar trends of conversion into countable entities are currently discussed for the representation of other fundamental units, e.g., the kilogram [178, 179].

The characterization of the obtained shot-noise suppression still seems to be limited by a drift-like increase of the PND. Interestingly, a theoretical treatment of the noise in mode-locked lasers predict constant power noise densities for feedback based CEP stabilizations [131]. Hence, this noise can only originate from effects that are not included in these treatments. The saturation of the power noise densities with decreasing frequency is also observed in recent measurements of the CEP error using a birefringence based f–$2f$ interferometer employing supercontinuum generation in an MSF [112], although pure interferometer noise measurements with the help of an continuous wave reference laser on a similar interferometer concept only exhibit a $1/f$-like increase without any saturation [111]. This indicates that interferometer drift may not be the reason for the observed increase of the power noise densities. As such a saturation is also not observed in measurements of the CEP error without spectral broadening in an MSF [73, 74] (see the negative slope of the power noise densities in Fig. 4.9), phase errors introduced inside the MSF, e.g., due to long-term

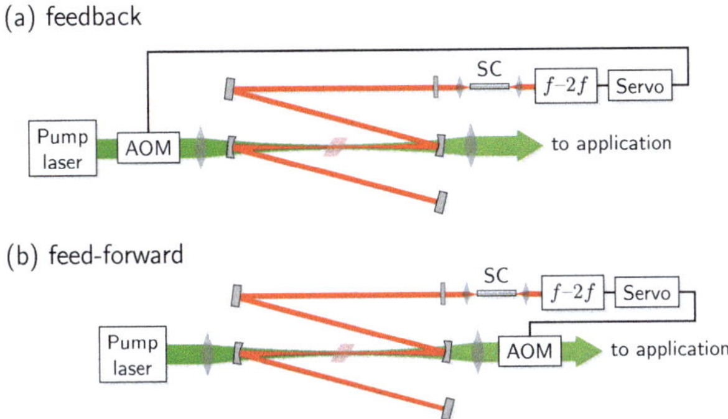

Figure 4.12.: (a) Feedback configuration and (b) feed-forward configuration for an out-of-loop characterization of the carrier-envelope frequency jitter. Here, it was assumed that the absorption process in the gain medium is employed as a beamsplitter for simplicity. However, feedback or feed-forward schemes may also be implemented by tapping off some fraction of the pump beam for applications before the detector. Alternatively, the known correlation of twin beams may be exploited by detecting the fluctuations of one of those and compensate for these fluctuations in the other twin beam.

variations of the energy coupled to the fiber, may be responsible for this feature of the power noise densities.

Outlook

For the clarification of the mechanism responsible for the sub-shot-noise power spectra, an out-of-loop characterization of the residual pump noise in "noise eater" configurations by means of analysis of the CEF jitter of a second free-running mode-locked laser is highly desired. Observation of sub-shot-noise CEF jitter of the out-of-loop mode-locked laser will give direct evidence of the presence of a QND-like conversion of pump power to CEF jitter. As depicted in Fig. 4.12, feedback or feed-forward "noise eater" configurations are possible. To the end of this ultimate demonstration, it is necessary to suppress external noise sources such as oscillator cavity length fluctuations and intracavity dispersion changes. A way to proceed in this direction may be to place the oscillator on a vibration decoupled bread board inside a vacuum tank. An alternative way may be to employ guided wave resonators such as fiber lasers do. Furthermore, approaching the predictions requires suppression of technical

RIN contributions of the pump laser. Given the sophistication of Ti:sapphire pump lasers, a further reduction of technical noise contributions appears to be limited if not even impossible. Thus, conversion to mode-locked lasers that are directly laser diode pumpable seems to be beneficial. As ultrashort pulse generation has already been shown for Yb-doped [180] and Cr:colquiriite [181] gain media, these crystals are promising alternatives. Furthermore, these crystals exhibit a low quantum defect of approximately 10% [180] and 15% [181], respectively. Hence, transition to these media may facilitate cooling and, therefore, lead to a reduction of thermal contributions to the per-roundtrip CEP shift.

In addition, such experiments will also benefit from a further increase of the sensitivity of the setups. In the above analysis, the sensitivity of the nonlinear phase shift φ_{nl} to pump power fluctuations has been identified as key parameter for the sensitivity of CEF towards pump power modulations. Detailed studies [95, 96] of the conversion of intracavity power into the per-roundtrip CEP shift show that this can be obtained in two ways. Holman et al. found that the per-roundtrip CEP shift is increased the higher the intracavity dispersion is and related this effect to a stronger shift of the carrier frequency [95]. In contrast, Sander et al. found that for octave spanning oscillators, i.e., for oscillators with negligible intracavity dispersion, there is a strong contribution of self-steepening [96]. First indications for the fact that broadband oscillators allow for a higher shot-noise reduction have been observed in the power noise densities extracted from the CEP stabilization measurements that have previously been published by other groups (Fig. 4.9). Additionally, improving the signal-to-noise ratio in the f–$2f$ interferometers and utilizing lasers with higher repetition rates increases the detectivity of the f–$2f$ front ends.

Furthermore, utilization of optical frequency combs generated in toroidal microcavities with terahertz repetition rates [182] may be worth investigating. Such frequency combs are generated via the interplay of the microcavity's resonances and four-wave mixing enabled by the Kerr nonlinearity of its host material. Hence, distinct from mode-locked lasers these frequency combs are generated without any intermediate absorption step. Comparison of results from microcavity experiments with experiments using mode-locked laser may, therefore, give valuable insight into which role the intermediate absorption plays within the conversion of pump power to CEP drift in mode-locked lasers.

In summary, the observation presented in this chapter calls for further enlightening experiments and bares the potential of generating strong photon-number squeezed light by analyzing the CEF of a mode-locked laser in combination with a feedback on the pump laser. Usage of such squeezed states are at hand for noise suppression in conventional light detection such as interferometry [183, 184], telecommunication [185], or spectroscopy [186].

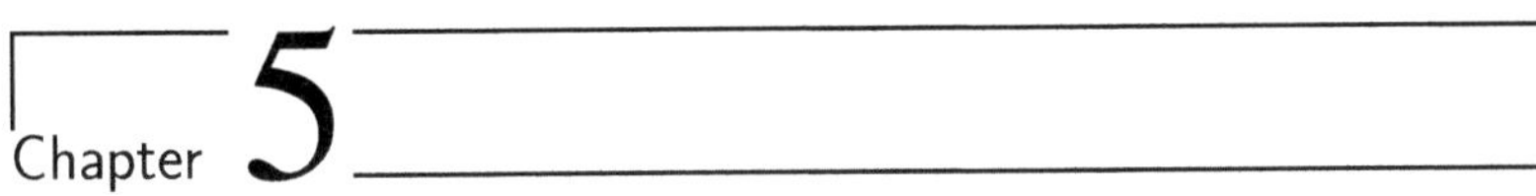

Chapter 5

Conclusions

The work presented in this thesis has successfully ascertained the origin of different carrier-envelope phase noise sources and showed that this jitter is due to both technical and fundamental noise generating mechanisms. In addition, various ways for elimination of these noise sources to the extent possible have been presented.

Carrier-envelope phase stabilization of oscillators was shown to be corrupted by spurious interferometer noise, which can easily be avoided by employing a common-path topology. Contrary to first expectations, the improvement does not chiefly stem from suppression of long-term drifts, but from a reduction of acoustic noise contributions. However, acoustic noise contributions could not be completely eliminated. This indicates that the phase-locked loop's bandwidth constraint, its amplitude-frequency characteristic, and the feedback to the laser are another source of phase error for the servo-loop approach. This finding is supported by the stabilization results obtained with a completely new approach to carrier-envelope phase stabilization, which is based on a feed-forward concept avoiding the shortcomings of the feedback-loop approach. Using this feed-forward concept, an unprecedented accuracy of the carrier-envelope phase stabilization is achieved with no observable noise contributions in the acoustic frequency region, being only limited by the influence of shot noise in the beat note detection. These results are not only interesting for the stabilization of few-cycle oscillators, but also for carrier-envelope phase stabilization of amplified laser systems used in attosecond physics. This is in particular true because the measurements with the two-detector phase extraction method revealed that the residual noise of the amplifier system is mainly inherited from the seed oscillator.

With these technical improvements at hand, a previously unrecognized fundamental limit of any carrier-envelope phase stabilization has been identified: the quantum noise statistics of the pump laser. This noise results in a hard limit for the achievable carrier-envelope phase jitter for frequencies below 100 Hz. Surprisingly, however, the carrier-envelope phase jitter that is dictated by the pump beam's quantum noise is found to be below the level expected for a pure conversion of the quantum noise into jitter of the carrier-envelope phase drift. This feature is explained by a feedback based squeezing and a quantum non-demolition like conversion of pump power to carrier-envelope phase jitter.

These results open a perspective towards the generation of more accurately carrier-envelope phase stabilized, high power pulses, by implementing the feed-forward concept into amplifier systems and combining it with the meaningful single-shot carrier-envelope phase extraction. Application of these pulses for attosecond pulse generation may help to further unravel the attosecond electron dynamics in atoms, molecules, and solids. If further experiments prove a dominant contribution of the quantum non-demolition like conversion of pump power to carrier-envelope phase drift, utilization of the sub-shot-noise performance of the carrier-envelope phase jitter of mode-locked lasers may enable the ultra-precise measurement of optical powers ranging from milliwatts, using microcavity-based frequency combs, to a few watts, employing mode-locked lasers. In addition, exploitation of the sub-shot-noise performance may open up an intriguing new path for sub-Poissonian light creation.

In summary, the work presented in this thesis paves an avenue towards advanced utilization of the carrier-envelope phase in existing and novel applications, enabling the study of uninvestigated physical phenomena.

Phase Noise Characterization

A.1. Frequency Analysis of the Carrier-Envelope Phase Noise

In order to characterize the accuracy of the CEP stabilization, the error signal of the CEP has been analyzed in the frequency domain. The detailed procedure of this analysis will be explained in this section.

Let us first recapitulate the basic definitions of Fourier analysis. Let $\varphi(t)$ denote the recorded time series of the CEP. By Fourier transforming the time series

$$\tilde{\varphi}(f) = \mathcal{F}\left[\varphi(t)\right] = \int\limits_{-\infty}^{\infty} \varphi(t)e^{-i2\pi ft}\,dt\,. \tag{A.1}$$

frequency contributions of the continuously varying phase error are evaluated. Since $\varphi(t)$ is measured only for a finite time period $N \cdot \Delta t$ and sampled with a rate of $1/\Delta t$ – which must not exceed the repetition rate of the laser – it is only possible to perform the discrete Fourier transform

$$\begin{aligned}
\tilde{\varphi}(f_k) &= \mathcal{DFT}\left[\varphi(t)\right] \\
&= \sum_{n=0}^{N-1} \varphi(t_n)e^{-i2\pi f_k t_n} \\
&= \sum_{n=0}^{N-1} \varphi(t_n)e^{-i2\pi kn/N}\,,
\end{aligned} \tag{A.2}$$

where $t_n = n\Delta t$ and $f_k = k\Delta f$. Here, it is assumed that the measurement started at time $t = 0$ and that $\Delta f = 1/(N\Delta t)$ denotes the lowest resolvable frequency, which decreases with increasing observation time. Since the phase $\varphi(t)$ is real-valued the spectrum is conjugate even, i.e., the relation $\tilde{\varphi}(f_k) = \tilde{\varphi}(f_{N-k})^*$ holds. Therefore, only one half of the array, e.g., bins $k = 0, ..., N/2$, contains non-redundant information. Since often only the strength of frequency f_k, i.e. the amplitude of $\tilde{\varphi}(f_k)$, is of interest one commonly applies the conversion

$$M_\varphi^{\mathrm{ss}}(f_k) = \begin{cases} |\tilde{\varphi}(f_0)|/N & \text{for } k = 0 \\ 2 \cdot |\tilde{\varphi}(f_k)|/N & \text{for } k = 1, \ldots, N/2 \end{cases} \tag{A.3}$$

to obtain a single-sided amplitude spectrum S^{os}. The values of S^{os} give the amplitude of each frequency bin in peak units. In order to convert from peak units to RMS units one has to divide the non-zero frequency components by the square-root of two

$$M_\varphi^{\mathrm{ss,rms}}(f_k) = \begin{cases} M_\varphi^{\mathrm{ss}}(f_0) & \text{for } k = 0 \\ M_\varphi^{\mathrm{ss}}(f_k)/\sqrt{2} & \text{for } k = 1, \ldots, N/2 \end{cases} \tag{A.4}$$

Normalizing the amplitude spectrum to the frequency resolution Δf

$$\begin{aligned} S_\varphi^{\mathrm{ss,rms}}(f_k) &= \frac{\left| M_\varphi^{\mathrm{ss,rms}}(f_k) \right|}{\sqrt{\Delta f}} \\ &= \begin{cases} |\tilde{\varphi}(f_k)|/(N\sqrt{\Delta f}) & \text{for } k = 0 \\ \sqrt{2} \cdot |\tilde{\varphi}(f_k)|/(N\sqrt{\Delta f}) & \text{for } k = 1, \ldots, N/2 \end{cases} \end{aligned} \tag{A.5}$$

renders the magnitude specification independent of the bandwidth of the measurement. The quantity defined in Eq. (A.5) is the so-called phase noise density (PND). The PND's independence from the measurement bandwidth is most important for measurements of noise levels where the frequency distribution is dense. Division by Δf yields a quantity that specifies the frequency contributions at frequency f normalized to the spectrum that would be measured if a 1 Hz wide square filter centered around f is applied before measurement. For convenience, the superscript is omitted throughout the thesis, and the single-sided, RMS-valued PND will be denoted by S_φ.

Due to Parseval's theorem, the integral

$$\delta_\varphi(f_m) = \sqrt{\sum_{k=m}^{N/2} \left(S_\varphi^{\mathrm{ss,rms}}(f_k) \right)^2 \Delta f} \,, \tag{A.6}$$

yields the phase error that is introduced by all noise contributions with frequencies greater than f_m. Hence, interpreting Eq. (A.6) in the time domain shows that the value of δ_φ at f_m corresponds to the phase error that is accumulated in time spans of length $1/f_m$.

Since any deviation of the phase evolution from a linear dependence also corresponds to a frequency variation ($f = d\varphi(t)/dt$), a PND can be converted into a frequency noise density (FND) S_f by multiplying with the Fourier frequency f [35, 141, 187]

$$S_f(f) = f S_\varphi(f) \,. \tag{A.7}$$

Often noise spectra exhibit strong fluctuations such that systematic features are hidden underneath. Averaging helps to isolate these systematic features from noise spectra. One elegant way to do so is the calculation of the PND by Welch's method: the time series under study is segmented into several, overlapping subsections. If not otherwise stated segmenting into eight sections with 50% overlap is applied throughout this thesis. In order to reduce biasing by edge effects a Hamming window

$$w_k = 0.54 - 0.46 \cos\left(2\pi k/N_{\mathrm{sub}}\right) \,, \quad 1 \le k \le N_{\mathrm{sub}} \tag{A.8}$$

is applied to the subsequences of length N_{sub}, and the PNDs of the individual subsections are averaged for estimation of the PND of the entire time series.

A.2. Phase Noise Measurements with a Radio-Frequency Spectrum Analyzer

The time domain representation $\varphi(t)$ has been the basis of the phase noise characterization method presented in the last section. Alternatively, phase noise of an RF oscillator can also be characterized by means of frequency domain measurement using a RF spectrum analyzer. As a prerequisite, the drift of the signal under test should not be larger than $\sim 1/10$ of the spectrum analyzer's span on the time span given by its sweep time.

Spectrum analyzers measure the signal's power in peak units at frequency f referenced to a sinusoidal signal at the same frequency having a power of 1 mW. For doing so the signal is heterodyned with a local oscillator, whose frequency is swept in time. The intermediate frequency obtained by this mixing is then low-pass filtered – this filter determines the resolution bandwidth (RBW) –, amplified and the power is detected with an envelope detector.

In consequence, spectrum analyzers measure the power spectrum. The intermediate frequency filter determines the noise levels observed in such measurements, while carrier levels remain constant under a change of the RBW. Thus like in the time domain characterization, the measured quantity has to be rendered insensitive to the RBW by normalizing it to the noise bandwidth B of the intermediate frequency filter, which can be found in the manufacturer's manual. Since the phase noise of an oscillation with carrier frequency f_c is to be characterized, the relevant frequency scale is the frequency difference $\Delta f = f - f_c$. After further division by the carrier signal level $L(f_c)$ and combined conversion of peak units to RMS units and to a single-sided spectrum (just consider $\Delta f \geq 0$), the single-sided phase noise density is obtained

$$S_\varphi^{\text{ss,rms}}(\Delta f) = \begin{cases} L(f_c + \Delta f)/(L(f_c) \cdot B) & \text{for } \Delta f = 0 \\ 2 \cdot L(f_c + \Delta f)/(L(f_c) \cdot B) & \text{for } \Delta f \neq 0 \end{cases} . \tag{A.9}$$

By inserting Eq. (A.9) into Eq. (A.6) for calculation of the integrated phase error one can easily deduce that for reducing the contribution of noise on the CEP jitter characterization by one order of magnitude requires a beat note with a $10^2 = 20$ dB better signal-to-noise ratio $L(f_c)/L(f_c + \Delta f)$.

Bibliography

[1] T. H. Maimann; *Stimulated Optical Radiation in Ruby*; Nature **187**, 493 (1960).

[2] R. J. Collins, D. F. Nelson, A. L. Schawlow, W. Bond, C. G. B. Garrett, and W. Kaiser; *Coherence, Narrowing, Directionality, and Relaxation Oscillations in the Light Emission from Ruby*; Phys. Rev. Lett. **5**, 303 (1960).

[3] A. H. Zewail; *Laser Femtochemistry*; Science **242**, 1645 (1988).

[4] M. DiDomenico, Jr.; *Small-Signal Analysis of Internal (Coupling-Type) Modulation of Lasers*; J. Appl. Phys. **35**, 2870 (1964).

[5] L. E. Hargrove, R. L. Fork, and M. A. Pollack; *Locking of He-Ne laser modes induced by synchronous intracavity modulation*; Appl. Phys. Lett. **5**, 4 (1964).

[6] S. Rausch, T. Binhammer, A. Harth, J. Kim, R. Ell, F. X. Kärtner, and U. Morgner; *Controlled waveforms on the single-cycle scale from a femtosecond oscillator*; Opt. Express **16**, 9739 (2008).

[7] A. de Bohan, P. Antoine, D. B. Milosević, and B. Piraux; *Phase-Dependent Harmonic Emission with Ultrashort Laser Pulses*; Phys. Rev. Lett. **81**, 1837 (1998).

[8] G. Tempea, M. Geissler, and T. Brabec; *Phase sensitivity of high-order harmonic generation with few-cycle laser pulses*; J. Opt. Soc. Am. B **16**, 669 (1999).

[9] E. Goulielmakis, M. Uiberacker, R. Kienberger, A. Baltuška, V. Yakovlev, A. Scrinzi, T. Westerwalbesloh, U. Kleineberg, U. Heinzmann, M. Drescher, and F. Krausz; *Direct Measurement of Light Waves*; Science **305**, 1267 (2004).

[10] J. N. Eckstein, A. I. Ferguson, and T. W. Hänsch; *High-Resolution Two-Photon Spectroscopy with Picosecond Light Pulses*; Phys. Rev. Lett. **40**, 847 (1978).

[11] H. Telle, G. Steinmeyer, A. Dunlop, J. Stenger, D. Sutter, and U. Keller; *Carrier-envelope offset phase control: A novel concept for absolute optical frequency measurement and ultrashort pulse generation*; Appl. Phys. B **69**, 327 (1999).

[12] J. Reichert, R. Holzwarth, T. Udem, and T. Hänsch; *Measuring the frequency of light with mode-locked lasers*; Opt. Comm. **172**, 59 (1999).

[13] T. Udem, R. Holzwarth, and T. W. Hänsch; *Optical frequency metrology*; Nature **416**, 233 (2002).

[14] S. A. Diddams, D. J. Jones, J. Ye, S. T. Cundiff, J. L. Hall, J. K. Ranka, R. S. Windeler, R. Holzwarth, T. Udem, and T. W. Hänsch; *Direct Link between Microwave and Optical Frequencies with a 300 THz Femtosecond Laser Comb*; Phys. Rev. Lett. **84**, 5102 (2000).

[15] R. Holzwarth, T. Udem, T. W. Hänsch, J. C. Knight, W. J. Wadsworth, and P. S. J. Russell; *Optical Frequency Synthesizer for Precision Spectroscopy*; Phys. Rev. Lett. **85**, 2264 (2000).

[16] H. Schnatz, B. Lipphardt, J. Helmcke, F. Riehle, and G. Zinner; *First Phase-Coherent Frequency Measurement of Visible Radiation*; Phys. Rev. Lett. **76**, 18 (1996).

[17] M. Fischer, N. Kolachevsky, M. Zimmermann, R. Holzwarth, T. Udem, T. W. Hänsch, M. Abgrall, J. Grünert, I. Maksimovic, S. Bize, H. Marion, F. P. D. Santos, P. Lemonde, G. Santarelli, P. Laurent, A. Clairon, C. Salomon, M. Haas, U. D. Jentschura, and C. H. Keitel; *New Limits on the Drift of Fundamental Constants from Laboratory Measurements*; Phys. Rev. Lett. **92**, 230802 (2004).

[18] O. Svelto; *Principles of Lasers* (Plenum Press, New York, 1998); ISBN 0-306-45748-2.

[19] A. J. DeMaria, D. A. Stetser, and H. Heynau; *Self mode-locking of lasers with saturable absorbers*; Appl. Phys. Lett. **8**, 174 (1966).

[20] U. Keller, D. A. B. Miller, G. D. Boyd, T. H. Chiu, J. F. Ferguson, and M. T. Asom; *Solid-state low-loss intracavity saturable absorber for Nd:YLF lasers: an antiresonant semiconductor Fabry–Perot saturable absorber*; Opt. Lett. **17**, 505 (1992).

[21] A. Schmidt, S. Rivier, G. Steinmeyer, J. H. Yim, W. B. Cho, S. Lee, F. Rotermund, M. C. Pujol, X. Mateos, M. Aguiló, F. Díaz, V. Petrov, and U. Griebner; *Passive mode locking of Yb:KLuW using a single-walled carbon nanotube saturable absorber*; Opt. Lett. **33**, 729 (2008).

[22] H. A. Haus; *Theory of mode locking with a fast saturable absorber*; J. Appl. Phys. **46**, 3049 (1975).

[23] H. Haus; *Theory of mode locking with a slow saturable absorber*; IEEE J. Quantum Electron. **11**, 736 (1975).

[24] V. Matsas, T. Newson, D. Richardson, and D. Payne; *Selfstarting passively mode-locked fibre ring soliton laser exploiting nonlinear polarisation rotation*; Electron. Lett. **28**, 1391 (1992).

[25] K. Tamura, H. Haus, and E. Ippen; *Self-starting additive pulse mode-locked erbium fibre ring laser*; Electron. Lett. **28**, 2226 (1992).

[26] E. P. Ippen, H. A. Haus, and L. Y. Liu; *Additive pulse mode locking*; J. Opt. Soc. Am. B **6**, 1736 (1989).

[27] P. N. Kean, X. Zhu, D. W. Crust, R. S. Grant, N. Langford, and W. Sibbett; *Enhanced mode locking of color-center lasers*; Opt. Lett. **14**, 39 (1989).

[28] D. E. Spence, P. N. Kean, and W. Sibbett; *60-fsec pulse generation from a self-mode-locked Ti:sapphire laser*; Opt. Lett. **16**, 42 (1991).

[29] F. Salin, J. Squier, and M. Piché; *Mode locking of Ti:Al$_2$O$_3$ lasers and self-focusing: a Gaussian approximation*; Opt. Lett. **16**, 1674 (1991).

[30] U. Keller, G. W. 'tHooft, W. H. Knox, and J. E. Cunningham; *Femtosecond pulses from a continuously self-starting passively mode-locked Ti:sapphire laser*; Opt. Lett. **16**, 1022 (1991).

[31] H. A. Haus, J. G. Fujimoto, and E. P. Ippen; *Structures for additive pulse mode locking*; J. Opt. Soc. Am. B **8**, 2068 (1991).

[32] T. Brabec, C. Spielmann, and F. Krausz; *Mode locking in solitary lasers*; Opt. Lett. **16**, 1961 (1991).

[33] Y. Chen, F. X. Kärtner, U. Morgner, S. H. Cho, H. A. Haus, E. P. Ippen, and J. G. Fujimoto; *Dispersion-managed mode locking*; J. Opt. Soc. Am. B **16**, 1999 (1999).

[34] R. Szipöcs, K. Ferencz, C. Spielmann, and F. Krausz; *Chirped multilayer coatings for broadband dispersion control in femtosecond lasers*; Opt. Lett. **19**, 201 (1994).

[35] F. W. Helbing, G. Steinmeyer, and U. Keller; *Carrier-Envelope Offset Phase-Locking With Attosecond Timing Jitter*; IEEE J. Sel. Top. Quantum Electron. **9**, 1030 (2003).

[36] F. W. Helbing, G. Steinmeyer, U. Keller, R. S. Windeler, J. Stenger, and H. R. Telle; *Carrier-envelope offset dynamics of mode-locked lasers*; Opt. Lett. **27**, 194 (2002).

[37] F. Helbing, G. Steinmeyer, J. Stenger, H. Telle, and U. Keller; *Carrier–envelope-offset dynamics and stabilization of femtosecond pulses*; Appl. Phys. B **74**, s35 (2002).

[38] T. Udem, J. Reichert, R. Holzwarth, and T. W. Hänsch; *Accurate measurement of large optical frequency differences with a mode-locked laser*; Opt. Lett. **24**, 881 (1999).

[39] S. A. Diddams, L. Hollberg, L.-S. Ma, and L. Robertsson; *Femtosecond-laser-based optical clockwork with instability $\leq 6.3 \times 10^{-16}$ in 1 s*; Opt. Lett. **27**, 58 (2002).

[40] L.-S. Ma, Z. Bi, A. Bartels, L. Robertsson, M. Zucco, R. S. Windeler, G. Wilpers, C. Oates, L. Hollberg, and S. A. Diddams; *Optical Frequency Synthesis and Comparison with Uncertainty at the 10-19 Level*; Science **303**, 1843 (2004).

[41] L. Hollberg, S. Diddams, A. Bartels, T. Fortier, and K. Kim; *The measurement of optical frequencies*; Metrologia **42**, S105 (2005).

[42] F. Krausz and M. Ivanov; *Attosecond physics*; Rev. Mod. Phys. **81**, 163 (2009).

[43] R. Kienberger, E. Goulielmakis, M. Uiberacker, A. Baltuška, V. Yakovlev, F. Bammer, A. Scrinzi, T. Westerwalbesloh, U. Kleineberg, U. Heinzmann, M. Drescher, and F. Krausz; *Atomic transient recorder*; Nature **427**, 817 (2004).

[44] E. Goulielmakis, M. Schultze, M. Hofstetter, V. S. Yakovlev, J. Gagnon, M. Uiberacker, A. L. Aquila, E. M. Gullikson, D. T. Attwood, R. Kienberger, F. Krausz, and U. Kleineberg; *Single-Cycle Nonlinear Optics*; Science **320**, 1614 (2008).

[45] I. J. Sola, E. Mevel, L. Elouga, E. Constant, V. Strelkov, L. Poletto, P. Villoresi, E. Benedetti, J.-P. Caumes, S. Stagira, C. Vozzi, G. Sansone, and M. Nisoli; *Controlling attosecond electron dynamics by phase-stabilized polarization gating*; Nat. Phys. **2**, 319 (2006).

[46] P. Tzallas, E. Skantzakis, C. Kalpouzos, E. P. Benis, G. D. Tsakiris, and D. Charalambidis; *Generation of intense continuum extreme-ultraviolet radiation by many-cycle laser fields*; Nat. Phys. **3**, 846 (2007).

[47] R. Trebino and D. J. Kane; *Using phase retrieval to measure the intensity and phase of ultrashort pulses: frequency-resolved optical gating*; J. Opt. Soc. Am. A **10**, 1101 (1993).

[48] C. Iaconis and A. Walmsley; *Spectral phase interferometry for direct electric-field reconstruction of ultrashort optical pulses*; Opt. Lett. **23**, 792 (1998).

[49] T. Brabec and F. Krausz; *Intense few-cycle laser fields: Frontiers of nonlinear optics*; Rev. Mod. Phys. **72**, 545 (2000).

[50] L. Xu, C. Spielmann, A. Poppe, T. Brabec, F. Krausz, and T. W. Hänsch; *Route to phase control of ultrashort light pulses*; Opt. Lett. **21**, 2008 (1996).

[51] S. T. Cundiff; *Phase stabilization of ultrashort optical pulses*; J. Phys. D **35**, R43 (2002).

[52] A. Baltuška, T. Fuji, and T. Kobayashi; *Controlling the Carrier-Envelope Phase of Ultrashort Light Pulses with Optical Parametric Amplifiers*; Phys. Rev. Lett. **88**, 133901 (2002).

[53] M. Kakehata, H. Takada, Y. Kobayashi, K. Torizuka, Y. Fujihira, T. Homma, and H. Takahashi; *Single-shot measurement of carrier-envelope phase changes by spectral interferometry*; Opt. Lett. **26**, 1436 (2001).

[54] T. Fuji, J. Rauschenberger, A. Apolonski, V. S. Yakovlev, G. Tempea, T. Udem, C. Gohle, T. W. Hänsch, W. Lehnert, M. Scherer, and F. Krausz; *Monolithic carrier-envelope phase-stabilization scheme*; Opt. Lett. **30**, 332 (2005).

[55] D. C. Heinecke, A. Bartels, T. M. Fortier, D. A. Braje, L. Hollberg, and S. A. Diddams; *Optical frequency stabilization of a 10 GHz Ti:sapphire frequency comb by saturated absorption spectroscopy in* ^{87}*rubidium*; Phys. Rev. A **80**, 053806 (2009).

[56] J. K. Ranka, R. S. Windeler, and A. J. Stentz; *Visible continuum generation in air silica microstructure optical fibers with anomalous dispersion at 800nm*; Opt. Lett. **25**, 25 (2000).

[57] R. R. Alfano and S. L. Shapiro; *Observation of Self-Phase Modulation and Small-Scale Filaments in Crystals and Glasses*; Phys. Rev. Lett. **24**, 592 (1970).

[58] D. J. Jones, S. A. Diddams, J. K. Ranka, A. Stentz, R. S. Windeler, J. L. Hall, and S. T. Cundiff; *Carrier-Envelope Phase Control of Femtosecond Mode-Locked Lasers and Direct Optical Frequency Synthesis*; Science **288**, 635 (2000).

[59] M. Mehendale, S. A. Mitchell, J. P. Likforman, D. M. Villeneuve, and P. B. Corkum; *Method for single-shot measurement of the carrier envelope phase of a few-cycle laser pulse*; Opt. Lett. **25**, 1672 (2000).

[60] G. Paulus, F. Lindner, H.Walther, A. Baltuška, E. Goulielmakis, M. Lezius, and F. Krausz; *Measurement of the Phase of Few-Cycle Laser Pulses*; Phys. Rev. Lett. **91**, 253004 (2003).

[61] C. A. Haworth, L. E. Chipperfield, J. S. Robinson, P. L. Knight, J. P. Marangos, and J. W. G. Tisch; *Half-cycle cutoffs in harmonic spectra and robust carrier-envelope phase retrieval*; Nat. Phys. **3**, 52 (2007).

[62] M. Kreß, T. Löffler, M. D. Thomson, R. Dörner, H. Gimpel, K. Zrost, T. Ergler, R. Moshammer, U. Morgner, J. Ullrich, and H. G. Roskos; *Determination of the carrier-envelope phase of few-cycle laser pulses with terahertz-emission spectroscopy*; Nat. Phys. **2**, 327 (2006).

[63] C. Lemell, X.-M. Tong, F. Krausz, and J. Burgdörfer; *Electron Emission from Metal Surfaces by Ultrashort Pulses: Determination of the Carrier-Envelope Phase*; Phys. Rev. Lett. **90**, 076403 (2003).

[64] A. Apolonski, P. Dombi, G. G. Paulus, M. Kakehata, R. Holzwarth, T. Udem, C. Lemell, K. Torizuka, J. Burgdörfer, T. W. Hänsch, and F. Krausz; *Observation of Light-Phase-Sensitive Photoemission from a Metal*; Phys. Rev. Lett. **92**, 073902 (2004).

[65] T. Wittmann, B. Horvath, W. Helml, M. G. Schatzel, X. Gu, A. L. Cavalieri, G. G. Paulus, and R. Kienberger; *Single-shot carrier-envelope phase measurement of few-cycle laser pulses*; Nat. Phys. **5**, 357 (2009).

[66] A. Apolonski, A. Poppe, G. Tempea, C. Spielmann, T. Udem, R. Holzwarth, T. W. Hänsch, and F. Krausz; *Controlling the Phase Evolution of Few-Cycle Light Pulses*; Phys. Rev. Lett. **85**, 740 (2000).

[67] A. Baltuška, T. Udem, M. Uiberacker, M. Hentschel, E. Goulielmakis, C. Gohle, R. Holzwarth, V. S. Yakovlev, A. Scrinzi, T. W. Hänsch, and F. Krausz; *Attosecond control of electronic processes by intense light fields*; Nature **421**, 611 (2003).

[68] L. Arissian and J.-C. Diels; *Carrier to envelope and dispersion control in a cavity with prism pairs*; Phys. Rev. A **75**, 013814 (2007).

[69] T. M. Fortier, D. J. Jones, J. Ye, S. T. Cundiff, and R. S. Windeler; *Long-term carrier-envelope phase coherence*; Opt. Lett. **27**, 1436 (2002).

[70] D. E. McCumber; *Intensity Fluctuations in the Output of cw Laser Oscillators. I*; Phys. Rev. **141**, 306 (1966).

[71] M. Sheik-Bahae, D. Hutchings, D. Hagan, and E. V. Stryland; *Dispersion of bound electron nonlinear refraction in solids*; IEEE J. Quantum Electron. **27**, 1296 (1991).

[72] L. Matos, O. D. Mücke, J. Chen, and F. X. Kärtner; *Carrier-envelope phase dynamics and noise analysis in octave-spanning Ti:sapphire lasers*; Opt. Express **14**, 2497 (2006).

[73] T. Fuji, J. Rauschenberger, C. Gohle, A. Apolonski, T. Udem, V. S. Yakovlev, G. Tempea, T. W. Hänsch, and F. Krausz; *Attosecond control of optical waveforms*; New J. Phys. **7**, 116 (2005).

[74] H. M. Crespo, J. R. Birge, E. L. Falcao-Filho, M. Y. Sander, A. Benedick, and F. X. Kärtner; *Nonintrusive phase stabilization of sub-two-cycle pulses from a prismless octave-spanning Ti:sapphire laser*; Opt. Lett. **33**, 833 (2008).

[75] J. Rauschenberger, T. Fuji, M. Hentschel, A.-J. Verhoef, T. Udem, C. Gohle, T. W. Hänsch, and F. Krausz; *Carrier-envelope phase-stabilized amplifier system*; Laser Phys. Lett. **3**, 37 (2006).

[76] C. Gohle, J. Rauschenberger, T. Fuji, T. Udem, A. Apolonski, F. Krausz, and T. W. Hänsch; *Carrier envelope phase noise in stabilized amplifier systems*; Opt. Lett. **30**, 2487 (2005).

[77] I. Thomann, E. Gagnon, R. Jones, A. Sandhu, A. Lytle, R. Anderson, J. Ye, M. Murnane, and H. Kapteyn; *Investigation of a grating-based stretcher/compressor for carrier-envelope phase stabilized fs pulses*; Opt. Express **12**, 3493 (2004).

[78] C. Li, E. Moon, H. Mashiko, C. M. Nakamura, P. Ranitovic, C. M. Maharjan, C. L. Cocke, Z. Chang, and G. G. Paulus; *Precision control of carrier-envelope phase in grating based chirped pulse amplifiers*; Opt. Express **14**, 11468 (2006).

[79] N. Forget, L. Canova, X. Chen, A. Jullien, and R. Lopez-Martens; *Closed-loop carrier-envelope phase stabilization with an acousto-optic programmable dispersive filter*; Opt. Lett. **34**, 3647 (2009).

[80] A. Baltuška, M. Uiberacker, E. Goulielmakis, R. Kienberger, V.Yakovlev, T. Udem, T. Hänsch, and F. Krausz; *Phase controlled amplification of few-cycle laser pulses*; IEEE J. Sel. Top. Quantum Electron. **9**, 972 (2003).

[81] G. Sansone, E. Benedetti, J.-P. Caumes, S. Stagira, C. Vozzi, S. De Silvestri, and M. Nisoli; *Control of long electron quantum paths in high-order harmonic generation by phase-stabilized light pulses*; Phys. Rev. A **73**, 053408 (2006).

[82] G. P. Agrawal; *Nonlinear Fiber Optics*; 2nd edition (Academic Press, San Diego, California, USA, 1995); ISBN 0-12-045142-5.

[83] J. C. Knight, T. A. Birks, P. S. J. Russell, and D. M. Atkin; *All-silica single-mode optical fiber with photonic crystal cladding*; Opt. Lett. **21**, 1547 (1996).

[84] T. A. Birks, W. J. Wadsworth, and P. S. J. Russell; *Supercontinuum generation in tapered fibers*; Opt. Lett. **25**, 1415 (2000).

[85] A. V. Husakou and J. Herrmann; *Supercontinuum Generation of Higher-Order Solitons by Fission in Photonic Crystal Fibers*; Phys. Rev. Lett. **87**, 203901 (2001).

[86] J. M. Dudley, G. Genty, and S. Coen; *Supercontinuum generation in photonic crystal fiber*; Rev. Mod. Phys. **78**, 1135 (2006).

[87] F. M. Mitschke and L. F. Mollenauer; *Discovery of the soliton self-frequency shift*; Opt. Lett. **11**, 659 (1986).

[88] J. P. Gordon; *Theory of the soliton self-frequency shift*; Opt. Lett. **11**, 662 (1986).

[89] W. Werncke, A. Lau, M. Pfeiffer, K. Lenz, H. J. Weigmann, and C. D. Thuy; *An anomalous frequency broadening in water*; Optics Communications **4**, 413 (1972).

[90] A. L. Gaeta; *Catastrophic Collapse of Ultrashort Pulses*; Phys. Rev. Lett. **84**, 3582 (2000).

[91] A. Brodeur and S. L. Chin; *Ultrafast white-light continuum generation and self-focusing in transparent condensed media*; J. Opt. Soc. Am. B **16**, 637 (1999).

[92] J. B. Ashcom, R. R. Gattass, C. B. Schaffer, and E. Mazur; *Numerical aperture dependence of damage and supercontinuum generation from femtosecond laser pulses in bulk fused silica*; J. Opt. Soc. Am. B **23**, 2317 (2006).

[93] M. Bradler, P. Baum, and E. Riedle; *Femtosecond continuum generation in bulk laser host materials with sub-μJ pump pulses*; Appl. Phys. B **97**, 561 (2009).

[94] Y. Yamamoto, S. Machida, and O. Nilsson; *Amplitude squeezing in a pump-noise-suppressed laser oscillator*; Phys. Rev. A **34**, 4025 (1986).

[95] K. Holman, R. Jones, A. Marian, S. Cundiff, and J. Ye; *Detailed studies and control of intensity-related dynamics of femtosecond frequency combs from mode-locked Ti:sapphire lasers*; IEEE J. Sel. Top. Quantum Electron. **9**, 1018 (2003).

[96] M. Y. Sander, E. P. Ippen, and F. X. Kärtner; *Carrier-envelope phase dynamics of octave-spanning dispersion-managed Ti:sapphire lasers*; Opt. Express **18**, 4948 (2010).

[97] K. L. Corwin, N. R. Newbury, J. M. Dudley, S. Coen, S. A. Diddams, K. Weber, and R. S. Windeler; *Fundamental Noise Limitations to Supercontinuum Generation in Microstructure Fiber*; Phys. Rev. Lett. **90**, 113904 (2003).

[98] T. M. Fortier, J. Ye, S. T. Cundiff, and R. S. Windeler; *Nonlinear phase noise generated in air-silica microstructure fiber and its effect on carrier-envelope phase*; Opt. Lett. **27**, 445 (2002).

[99] N. Haverkamp and H. Telle; *Complex intensity modulation transfer function for supercontinuum generation in microstructure fibers*; Opt. Express **12**, 582 (2004).

[100] B. Washburn and N. Newbury; *Phase, timing, and amplitude noise on supercontinua generated in microstructure fiber*; Opt. Express **12**, 2166 (2004).

[101] S. Koke, C. Grebing, B. Manschwetus, and G. Steinmeyer; *Fast f-to-2f interferometer for a direct measurement of the carrier-envelope phase drift of ultrashort amplified laser pulses*; Opt. Lett. **33**, 2545 (2008).

[102] C. Grebing, S. Koke, B. Manschwetus, and G. Steinmeyer; *Performance comparison of interferometer topologies for carrier-envelope phase detection*; Appl. Phys. B **95**, 81 (2009).

[103] C. Grebing; *Neuartige Konzepte zur Detektion und Kontrolle der Carrier-Envelope Phasendrift ultrakurzer Laserimpulse*; Ph.D. thesis; Humboldt-Universität zu Berlin (2010).

[104] T. M. Fortier, A. Bartels, and S. A. Diddams; *Octave-spanning Ti:sapphire laser with a repetition rate >1 GHz for optical frequency measurements and comparisons*; Opt. Lett. **31**, 1011 (2006).

[105] E. Moon, C. Li, Z. Duan, J. Tackett, K. L. Corwin, B. R. Washburn, and Z. Chang; *Reduction of fast carrier-envelope phase jitter in femtosecond laser amplifiers*; Opt. Express **14**, 9758 (2006).

[106] U. Morgner, R. Ell, G. Metzler, T. R. Schibli, F. X. Kärtner, J. G. Fujimoto, H. A. Haus, and E. P. Ippen; *Nonlinear Optics with Phase-Controlled Pulses in the Sub-Two-Cycle Regime*; Phys. Rev. Lett. **86**, 5462 (2001).

[107] T. R. Schibli, K. Minoshima, F.-L. Hong, H. Inaba, A. Onae, H. Matsumoto, I. Hartl, and M. E. Fermann; *Frequency metrology with a turnkey all-fiber system*; Opt. Lett. **29**, 2467 (2004).

[108] J. J. McFerran, W. C. Swann, B. R. Washburn, and N. R. Newbury; *Elimination of pump-induced frequency jitter on fiber-laser frequency combs*; Opt. Lett. **31**, 1997 (2006).

[109] F. W. Helbing; *Measurement and Control of the Carrier-Envelope Offset Phase*; Ph.D. thesis; Swiss Federal Institute of Technology Zurich (2004).

[110] D. J. Jones, T. M. Fortier, and S. T. Cundiff; *Highly sensitive detection of the carrier-envelope phase evolution and offset of femtosecond mode-locked oscillators*; J. Opt. Soc. Am. B **21**, 1098 (2004).

[111] M. Pawlowska, F. Ozimek, P. Fita, and C. Radzewicz; *Collinear interferometer with variable delay for carrier-envelope offset frequency measurement*; Rev. Sci. Instrum. **80**, 083101 (2009).

[112] V. Tsatourian, H. Margolis, G. Marra, D. Reid, and P. Gill; *Common-path self-referencing interferometer for carrier-envelope offset frequency stabilization with enhanced noise immunity*; Opt. Lett. **35**, 1209 (2010).

[113] A. Guandalini, P. Eckle, M. Anscombe, P. Schlup, J. Biegert, and U. Keller; *5.1 fs pulses generated by filamentation and carrier envelope phase stability analysis*; J. Phys. B **39**, S257 (2006).

[114] A. M. Sayler, T. Rathje, W. Müller, K. Rühle, R. Kienberger, and G. G. Paulus; *Precise, real-time, every-single-shot, carrier-envelope phase measurement of ultrashort laser pulses*; Opt. Lett. **36**, 1 (2011).

[115] H. Wang, M. Chini, E. Moon, H. Mashiko, C. Li, and Z. Chang; *Coupling between energy and phase in hollow-core fiber based f-to-2f interferometers*; Opt. Express **17**, 12082 (2009).

[116] H. M. Crespo, J. R. Birge, M. Y. Sander, E. L. Falcão Filho, A. Benedick, and F. X. Kärtner; *Phase stabilization of sub-two-cycle pulses from prismless octave-spanning Ti:sapphire lasers*; J. Opt. Soc. Am. B **25**, B147 (2008).

[117] S. Rausch, T. Binhammer, A. Harth, E. Schulz, M. Siegel, and U. Morgner; *Few-cycle oscillator pulse train with constant carrier-envelope- phase and 65 as jitter*; Opt. Express **17**, 20282 (2009).

[118] S. Koke, C. Grebing, H. Frei, A. Anderson, A. Assion, and G. Steinmeyer; *Direct frequency comb synthesis with arbitrary offset and shot-noise-limited phase noise*; Nat. Photon. **4**, 462 (2010).

[119] R. W. Dixon; *Photoelastic Properties of Selected Materials and Their Relevance for Applications to Acoustic Light Modulators and Scanners*; J. Appl. Phys. **38**, 5149 (1967).

[120] A. Korpel; *Acousto-Optics* (Marcel Dekker, Inc., New York, 1988); ISBN 0-8247-7891-X.

[121] B. E. A. Saleh and M. C. Teich; *Fundamentals of Photonics* (John Wiley & Sons, New York, 1991); ISBN 978-0-471-35832-9.

[122] Y. S. Lee, J. Sung, C. Nam, T. Yu, and K.-H. Hong; *Novel method for carrier-envelope-phase stabilization of femtosecond laser pulses*; Opt. Express **13**, 2969 (2005).

[123] S. Witte, R. Zinkstok, W. Hogervorst, and K. Eikema; *Control and precise measurement of carrier-envelope phase dynamics*; Appl. Phys. B **78**, 5 (2004).

[124] C. Vozzi, C. Manzoni, F. Calegari, E. Benedetti, G. Sansone, G. Cerullo, M. Nisoli, S. De Silvestri, and S. Stagira; *Characterization of a high-energy self-phase-stabilized near-infrared parametric source*; J. Opt. Soc. Am. B **25**, B112 (2008).

[125] A. Renault, D. Z. Kandula, S. Witte, A. L. Wolf, R. T. Zinkstok, W. Hogervorst, and K. S. E. Eikema; *Phase stability of terawatt-class ultrabroadband parametric amplification*; Opt. Lett. **32**, 2363 (2007).

[126] M. Takeda, H. Ina, and S. Kobayashi; *Fourier-transform method of fringe-pattern analysis for computer-based topography and interferometry*; J. Opt. Soc. Am. **72**, 156 (1982).

[127] E. Benkler, H. R. Telle, K. Weingarten, L. Krainer, G. Spühler, and U. Keller; *Characterization of ultrashort optical pulse properties by amplitude-modulation-balanced heterodyne gating*; Opt. Lett. **30**, 2016 (2005).

[128] H. Telle, B. Lipphardt, and J. Stenger; *Kerr-lens, mode-locked lasers as transfer oscillators for optical frequency measurements*; Appl. Phys. B **74**, 1 (2002).

[129] A. Fernandez, T. Fuji, A. Poppe, A. Fürbach, F. Krausz, and A. Apolonski; *Chirped-pulse oscillators: a route to high-power femtosecond pulses without external amplification*; Opt. Lett. **29**, 1366 (2004).

[130] F. X. Kärtner, U. Morgner, T. Schibli, R. Ell, H. A. Haus, J. G. Fujimoto, and E. P. Ippen; *Few-Cycle Laser Pulse Generation and Its Applications*; chapter Few-Cycle Pulses Directly from a Laser, 73–136 (Springer, Berlin, 2004); ISBN 3-540-20115-7.

[131] J. K. Wahlstrand, J. T. Willits, C. R. Menyuk, and S. T. Cundiff; *The quantum-limited comb lineshape of a mode-locked laser: Fundamental limitson frequency uncertainty*; Opt. Express **16**, 18624 (2008).

[132] Y. Yamamoto, N. Imoto, and S. Machida; *Amplitude squeezing in a semiconductor laser using quantum nondemolition measurement and negative feedback*; Phys. Rev. A **33**, 3243 (1986).

[133] H. A. Haus and Y. Yamamoto; *Theory of feedback-generated squeezed states*; Phys. Rev. A **34**, 270 (1986).

[134] H. M. Wiseman; *Quantum Squeezing*; chapter Squeezing and Feedback, 171–223 (Springer, Berlin, 2004); ISBN 978-3-540-65989-1.

[135] A. Einstein; *Über einen die Erzeugung und Verwandlung des Lichtes betreffenden heuristischen Gesichtspunkt*; Annalen der Physik **322**, 132 (1905).

[136] J. F. Clauser; *Experimental distinction between the quantum and classical field-theoretic predictions for the photoelectric effect*; Phys. Rev. D **9**, 853 (1974).

[137] R. Loudon; *Non-classical effects in the statistical properties of light*; Rep. Progr. Phys. **43**, 913 (1980).

[138] D. F. Walls; *Evidence for the quantum nature of light*; Nature **280**, 451 (1979).

[139] N. Campbell; *The study of discontinuous phenomena. Discontinuities in light emission*; Proc. Cambr. Phil. Soc. **15**, 310 (1909).

[140] O. Rice; *Mathematical Analysis of Random Noise*; Bell Syst. Tech. J. **23**, 282 (1944).

[141] E. Ferre-Pikal, J. Vig, J. Camparo, L. Cutler, L. Maleki, W. Riley, S. Stein, C. Thomas, F. Walls, and J. White; *Draft revision of IEEE STD 1139-1988 standard definitions of physical quantities for fundamental, frequency and time metrology-random instabilities*; in *Frequency Control Symposium, 1997*; 338–357 (1997).

[142] W. Riley; *Handbook of Frequency Stability Analysis* (NIST SP - 1065, 2008).

[143] J. Mertz, A. Heidmann, C. Fabre, E. Giacobino, and S. Reynaud; *Observation of high-intensity sub-Poissonian light using an optical parametric oscillator*; Phys. Rev. Lett. **64**, 2897 (1990).

[144] D. F. Walls; *Squeezed states of light*; Nature **306**, 141 (1983).

[145] G. Breitenbach, S. Schiller, and J. Mlynek; *Measurement of the quantum states of squeezed light*; Nature **387**, 471 (1997).

[146] S. R. Friberg, S. Machida, M. J. Werner, A. Levanon, and T. Mukai; *Observation of Optical Soliton Photon-Number Squeezing*; Phys. Rev. Lett. **77**, 3775 (1996).

[147] S. Schmitt, J. Ficker, M. Wolff, F. König, A. Sizmann, and G. Leuchs; *Photon-Number Squeezed Solitons from an Asymmetric Fiber-Optic Sagnac Interferometer*; Phys. Rev. Lett. **81**, 2446 (1998).

[148] W. H. Richardson, S. Machida, and Y. Yamamoto; *Squeezed photon-number noise and sub-Poissonian electrical partition noise in a semiconductor laser*; Phys. Rev. Lett. **66**, 2867 (1991).

[149] Y. Yamamoto, S. Machida, and W. H. Richardson; *Photon Number Squeezed States in Semiconductor Lasers*; Science **255**, 1219 (1992).

[150] S. Machida and Y. Yamamoto; *Observation of sub-poissonian photoelectron statistics in a negative feedback semiconductor laser*; Opt. Comm. **57**, 290 (1986).

[151] S. R. Friberg and S. Machida; *Ultrafast optical pulse noise suppression using a nonlinear spectral filter: 23 dB reduction of fiber laser 1/f noise*; Appl. Phys. Lett. **73**, 1934 (1998).

[152] F. X. Kärtner and H. A. Haus; *Quantum-nondemolition measurements and the "collapse of the wave function"*; Phys. Rev. A **47**, 4585 (1993).

[153] V. B. Braginsky, Y. I. Vorontsov, and K. S. Thorne; *Quantum Nondemolition Measurements*; Science **209**, 547 (1980).

[154] C. M. Caves, K. S. Thorne, R. W. P. Drever, V. D. Sandberg, and M. Zimmermann; *On the measurement of a weak classical force coupled to a quantum-mechanical oscillator. I. Issues of principle*; Rev. Mod. Phys. **52**, 341 (1980).

[155] P. Grangier, J. A. Levenson, and J.-P. Poizat; *Quantum non-demolition measurements in optics*; Nature **396**, 537 (1998).

[156] J.-P. Poizat, J.-F. Roch, and P. Grangier; *Characterization of quantum non-demolition measurements in optics*; Ann. Phys. Fr. **19**, 265 (1994).

[157] N. Imoto, H. A. Haus, and Y. Yamamoto; *Quantum nondemolition measurement of the photon number via the optical Kerr effect*; Phys. Rev. A **32**, 2287 (1985).

[158] J. Weber; *Gravitational-Wave-Detector Events*; Phys. Rev. Lett. **20**, 1307 (1968).

[159] H. A. Haus; *Electromagnetic Noise and Quantum Optical Measurements* (Springer-Verlag, Berlin, 2000); ISBN 3-540-65272-8.

[160] R. J. Glauber; *Coherent and Incoherent States of the Radiation Field*; Phys. Rev. **131**, 2766 (1963).

[161] D. F. Walls and G. J. Milburn; *Quantum optics* (Springer-Verlag, Berlin, 1994); ISBN 3-540-57179-5.

[162] Y. Yamamoto and H. A. Haus; *Preparation, measurement and information capacity of optical quantum states*; Rev. Mod. Phys. **58**, 1001 (1986).

[163] S. R. Friberg, S. Machida, and Y. Yamamoto; *Quantum-nondemolition measurement of the photon number of an optical soliton*; Phys. Rev. Lett. **69**, 3165 (1992).

[164] P. D. Drummond, R. M. Shelby, S. R. Friberg, and Y. Yamamoto; *Quantum solitons in optical fibres*; Nature **365**, 307 (1993).

[165] P. Grangier, J.-F. Roch, and G. Roger; *Observation of backaction-evading measurement of an optical intensity in a three-level atomic nonlinear system*; Phys. Rev. Lett. **66**, 1418 (1991).

[166] J. F. Roch, G. Roger, P. Grangier, J.-M. Courty, and S. Reynaud; *Quantum non-demolition measurements in optics: a review and some recent experimental results*; Appl. Phys. B **55**, 291 (1992).

[167] J. P. Poizat and P. Grangier; *Experimental realization of a quantum optical tap*; Phys. Rev. Lett. **70**, 271 (1993).

[168] J.-F. Roch, K. Vigneron, P. Grelu, A. Sinatra, J.-P. Poizat, and P. Grangier; *Quantum Nondemolition Measurements using Cold Trapped Atoms*; Phys. Rev. Lett. **78**, 634 (1997).

[169] M. D. Levenson, R. M. Shelby, M. Reid, and D. F. Walls; *Quantum Nondemolition Detection of Optical Quadrature Amplitudes*; Phys. Rev. Lett. **57**, 2473 (1986).

[170] A. La Porta, R. E. Slusher, and B. Yurke; *Back-Action Evading Measurements of an Optical Field Using Parametric Down Conversion*; Phys. Rev. Lett. **62**, 28 (1989).

[171] S. F. Pereira, Z. Y. Ou, and H. J. Kimble; *Backaction evading measurements for quantum nondemolition detection and quantum optical tapping*; Phys. Rev. Lett. **72**, 214 (1994).

[172] R. Bruckmeier, K. Schneider, S. Schiller, and J. Mlynek; *Quantum Nondemolition Measurements Improved by a Squeezed Meter Input*; Phys. Rev. Lett. **78**, 1243 (1997).

[173] K. Bencheikh, J. A. Levenson, P. Grangier, and O. Lopez; *Quantum Nondemolition Demonstration via Repeated Backaction Evading Measurements*; Phys. Rev. Lett. **75**, 3422 (1995).

[174] R. Bruckmeier, H. Hansen, and S. Schiller; *Repeated Quantum Nondemolition Measurements of Continuous Optical Waves*; Phys. Rev. Lett. **79**, 1463 (1997).

[175] A. Abramovici, W. E. Althouse, R. W. P. Drever, Y. Gursel, S. Kawamura, F. J. Raab, D. Shoemaker, L. Sievers, R. E. Spero, K. S. Thorne, R. E. Vogt, R. Weiss, S. E. Whitcomb, and M. E. Zucker; *LIGO: The Laser Interferometer Gravitational-Wave Observatory*; Science **256**, 325 (1992).

[176] M. Ando, K. Arai, R. Takahashi, G. Heinzel, S. Kawamura, D. Tatsumi, N. Kanda, H. Tagoshi, A. Araya, H. Asada, Y. Aso, M. A. Barton, M.-K. Fujimoto, M. Fukushima, T. Futamase, K. Hayama, G. Horikoshi, H. Ishizuka, N. Kamikubota, K. Kawabe, N. Kawashima, Y. Kobayashi, Y. Kojima, K. Kondo, Y. Kozai, K. Kuroda, and N. Matsuda; *Stable Operation of a 300-m Laser Interferometer with Sufficient Sensitivity to Detect Gravitational-Wave Events within Our Galaxy*; Phys. Rev. Lett. **86**, 3950 (2001).

[177] T. W. Hänsch; *Nobel Lecture: Passion for precision*; Rev. Mod. Phys. **78**, 1297 (2006).

[178] J. Flowers; *The Route to Atomic and Quantum Standards*; Science **306**, 1324 (2004).

[179] P. Becker; *Tracing the definition of the kilogram to the Avogadro constant using a silicon single crystal*; Metrologia **40**, 366 (2003).

[180] H. Liu, J. Nees, and G. Mourou; *Diode-pumped Kerr-lens mode-locked Yb:KY(WO$_4$)$_2$ laser*; Opt. Lett. **26**, 1723 (2001).

[181] P. C. Wagenblast, U. Morgner, F. Grawert, T. R. Schibli, F. X. Kärtner, V. Scheuer, G. Angelow, and M. J. Lederer; *Generation of sub-10-fs pulses from a Kerr-lens mode-locked Cr^{3+}:LiCAF laser oscillator by use of third-order dispersion-compensating double-chirped mirrors*; Opt. Lett. **27**, 1726 (2002).

[182] P. Del'Haye, A. Schliesser, O. Arcizet, T. Wilken, R. Holzwarth, and T. J. Kippenberg; *Optical frequency comb generation from a monolithic microresonator*; Nature **450**, 1214 (2007).

[183] M. Xiao, L.-A. Wu, and H. J. Kimble; *Precision measurement beyond the shot-noise limit*; Phys. Rev. Lett. **59**, 278 (1987).

[184] P. Grangier, R. E. Slusher, B. Yurke, and A. La Porta; *Squeezed-light-enhanced polarization interferometer*; Phys. Rev. Lett. **59**, 2153 (1987).

[185] B. E. A. Saleh and M. C. Teich; *Can the channel capacity of a light-wave communication system be increased by the use of photon-number squeezed light?*; Phys. Rev. Lett. **58**, 2656 (1987).

[186] E. S. Polzik, J. Carri, and H. J. Kimble; *Spectroscopy with squeezed light*; Phys. Rev. Lett. **68**, 3020 (1992).

[187] T. Fortier, D. Jones, J. Ye, and S. Cundiff; *Highly phase stable mode-locked lasers*; IEEE J. Sel. Top. Quantum Electron. **9**, 1002 (2003).

Danksagung

Mein besonderer Dank gilt Prof. Dr. Thomas Elsässer für die Betreuung meiner Promotion und die Möglichkeit, meine Forschung auf internationalen Konferenzen zu präsentieren.

PD Dr. Günter Steinmeyer danke ich für die Weitergabe seines umfangreichen Wissens über Ultrakurzpulslaser, die durchgehende Betreuung meiner Arbeit sowie seine Beharrlichkeit, die letzten Endes zu den hochrangig veröffentlichten Ergebnissen geführt hat. Bei Dr. Christian Grebing, Jens Bethge, Dr. Thomas Hansel, Bastian Borchers, Carsten Brée, Andreas Schmidt, Martin Bock und Simon Birkholz bedanke ich mich für ihre Kollegialität und die ausgiebigen Diskussionen.

Mein Dank gilt auch dem Team von Femtolasers, insbesondere PD Dr. Andreas Assion, Harald Frei und Alexandria Anderson, für die Kooperationsbereitschaft und die Unterstützung meiner Messungen in Wien. Darüber hinaus möchte ich Prof. Dr. Roman Schnabel für Diskussionen zu dem letzten Kapitel herzlich danken. Des Weiteren danke ich Dr. Olga Smirnova, Prof. Dr. Martin Weinelt und Prof. Dr. Claus Ropers für die aufschlußreichen Diskussionen über die Carrier-Envelope-Phasenabhängigkeit der Elektronenemission aus Metallen, auch wenn dieses Thema am Ende nicht in die Arbeit aufgenommen wurde.

Bedanken möchte ich mich außerdem bei der gesamten Abteilung C2 für die überaus freundliche und motivierende Arbeitsatmosphäre. Insbesondere gilt mein Dank Dorit Fischer für die Vermittlung ihrer umfangreichen Erfahrungen in der Präparation von optischen Fasern sowie Wolfgang Goleschny für die Konstruktion und Umsetzung mechanischer Komponenten. Meinem Büronachbarn Dr. Roland Müller danke ich für das angenehme Miteinander, die interessanten Gespräche und seine stete Hilfsbereitschaft.

Zu guter Letzt möchte ich mich von ganzem Herzen bei meiner Frau für ihre unermüdliche Unterstützung und ihr Verständnis bedanken.

Selbständigkeitserklärung

Hiermit erkläre ich, dass ich die vorliegende Dissertation selbständig und nur unter Verwendung der angegebenen Hilfen und Hilfsmittel angefertigt habe.

Berlin, den 24. März 2011   Sebastian Koke